STRATEGY AND TACTICS:

BY

GENERAL G. H. DUFOUR,

LATELY AN OFFICER OF THE FRENCH ENGINEER CORPS, GRADUATE OF THE POLYTECHNIC SCHOOL, AND COMMANDER OF THE LEGION OF HONOR; CHIEF OF STAFF OF THE SWISS ARMY.

TRANSLATED FROM THE LATEST FRENCH EDITION,

BY

WM. P. CRAIGHILL,

CAPT. U. S. ENGINEERS, LATELY ASST. PROF. OF CIVIL AND MILITARY ENGINEERING AND SCIENCE OF WAR AT THE U. S. MILITARY ACADEMY.

NEW YORK:
D. VAN NOSTRAND, 192 BROADWAY.
1864.

C. A. ALVORD, STEREOTYPER AND PRINTER.

TRANSLATOR'S PREFACE.

General Dufour, the Chief of the General Staff of the Army of Switzerland, is a graduate of the Polytechnic School of France,—the institution styled by the great Napoleon, "*The hen which lays me golden eggs.*" General Dufour is a distinguished Civil and Military Engineer, besides being a practical soldier. In all military matters he is recognized as one of the first authorities in Europe.

The work on Strategy, Guard Tactics, &c., of which a translation is now offered to the public, was prepared for the instruction of the officers of the Swiss army. It is written in a plain and simple style. The author does not deal in vague generalities, but gives simple, practical directions, which are illustrated by actual examples left by Napoleon and other great generals of ancient and modern times. It will be observed that the work contains numerous excellent plates embodying principles in themselves.

That portion of the original which had a special application to the Swiss army has been omitted in the translation. Some modifications and slight additions have been ventured upon, with a view of making the book more acceptable to American readers.

U. S. MILITARY ACADEMY,
West Point, N. Y.

PLATES.

PRINCIPLES OF STRATEGY.

CHAPTER I.

Art. I.—Definitions.

If by the aid of the histories, often very incomplete, of the armies of ancient times, we examine their operations, we shall find some of them more or less similar to those of modern armies, and others essentially different. The first class, which relate to grand movements, are within the domain of *Strategy.* The second class belong to Tactics: such are the dispositions for marches, for battles, of camps, &c. For example, the famous expeditions of Hannibal and Napoleon across the Alps presented more than one point of resemblance in their general outline, but the subsequent combats and battles were entirely dissimilar. The general arrangements of campaigns depend upon the topography of the theatre of the war, which is almost unchanging; but special arrangements, evolutions, or, in a word, Tactics, vary with the arms in use at different periods. Much valuable instruction in *strategy* may therefore be derived from the study of history: but very grave errors would result if we attempt to apply in the armies of the present day the *tactics* of the ancients. This fault has been committed by more than one man of merit, for want of reflection upon the great differences between our missile-weapons and those of the ancients, and upon the resulting differences in the arrangement of troops for combat.

It is evident from the preceding remarks, that strategy is in a special manner *the science of generals;* but in tactics, in its various ramifications, from the school of the platoon to the orders of battle, from the bivouac of an outpost to the camping of an army, officers of all grades are concerned. It would, however, be pedantry to desire to fix invariable limits for these two branches of the military art, for there are many cases where these separating limits disappear.

Tactics, if we designate by that term simply the evolutions of the battle-field, may, so to speak, be taught by mathematical figures, because for each of these movements there is a rule; and this remark is the more applicable as we descend the scale of military knowledge. But the case is not the same with strategy, because in its calculations account must be taken of time, of the nature of roads, of obstacles of every kind, of the quality of troops, of the mobility of the enemy, and many other things which cannot be measured by mathematical instruments. For example, any one would be greatly deceived who should believe the point A (fig. 1) sufficiently protected by the army M against an enemy placed at N, a little outside the circle described with A as a centre and A M as a radius. Athough, upon the supposition that the ground is perfectly level, the army M may reach A before the army N, which is at a greater distance; still the point A is insecure, because the enemy, taking the initiative before the army M is informed of its movement, may have so much the start as to reach some point, as P, before M is ready to march; the distance P A being then less than M A, N could arrive at A before M. It is quite evident from this illustration, that other things besides distances are to be taken into account in calculating the relative advantages of the positions of two armies having a common object in view: the time must be considered which is necessary for one army to be informed of the movement of the other; and this element of time is the one which in war gives so many advantages to the general who takes the initiative, and moves with activity. If, moreover, the quality of the roads is variable; if rivers are to be crossed and defiles to be passed—as the two armies will seldom be upon equal terms in these particulars—the problem

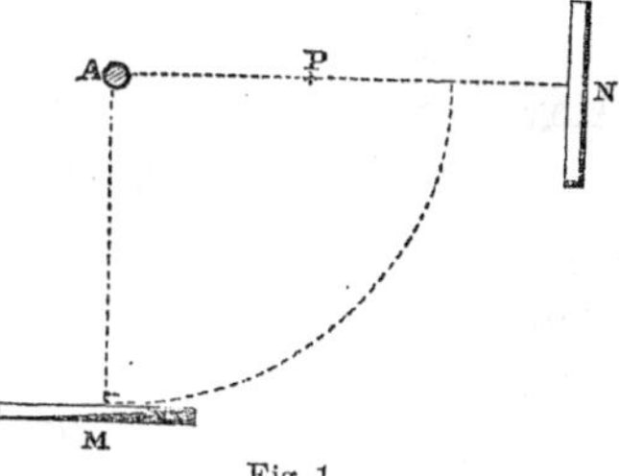

Fig. 1.

of the relative advantages of positions becomes still more complicated, and less capable of solution by the use of geometrical figures; genius alone can grasp and solve it.

Having now explained what is to be understood by the terms *strategy* and *tactics*, to the definition of which too much importance has perhaps been lately given, we shall proceed with our subject.

In every operation in war there are three principal things to be considered: the point of departure, the point to be attained, and the direction to be followed in reaching it. The line upon which an army rests, and from which it proceeds to act against the enemy, is called the *base of operations*. Thus, the Rhine, with its fortified cities, is the base of operations for the French in waging a war in Germany. The Alps and Pyrenees, with their forts and defiles, are the bases for operations in Italy and Spain.

If the army remains upon its base of operations, and limits itself to disputing its possession with the enemy, it takes the name of *line of defence*. In following the course of the Danube, we find in succession the lines of defence of the Black Forest; of the Iller, which has Ulm on the right; of the Lech, passing by Augsburg; of the Iser, having Munich and Landshut near it; of the Inn, which empties into the Danube at Passau, &c. A hundred combats have taken place on these different lines.

The points which it is desirable to attain, whose

capture is important for the success of a campaign, are called *objective points*, or *objectives of operations.*

The route pursued by an army in reaching an *objective* is the *line of operations*, which takes the name of *line of retreat* when passed over by an army retiring before a victorious enemy, in order to gain some line of defence where a stand may be made.

Art. II.—Bases of Operations.

The base of operations should be formed of secure points, where every thing is collected that is necessary for the wants of the army, and from which proceed roads suitable for purposes of transportation. If these points are joined by a natural obstacle, such as a river, a chain of mountains, a great marsh, extensive forests, the line is so much the better, on account of the difficulties presented to an enemy who attempts to break through it.

A base of operations which has some extent is better than a contracted one, because there is more room for manœuvres, and there is less danger of being cut off from it. If, for example, a single city forms the base, and an enemy gets possession of the principal road leading to it, the army may be greatly embarrassed, as it will be cut off from all supplies and re-enforcements.

The form of the base is not a matter of indifference: if it is concave to the front, or has its extremities resting upon the sea, a great lake, or some other prominent obstacle, the army has its wings much more secure than if the base be convex, or presents a point towards the enemy.

When an army moves forward a long distance, it is obliged to assume a new base in advance of the first, in order to have dépôts at hand from which its supplies may be drawn. This new line is called the *secondary base of operations;* it is usually a river crossing the line of operations, and the towns upon it are fortified, if they are not already so, in order to the security of the military supplies, &c., collected in them.

If it is prudent to occupy successive bases of operations as the army advances into an enemy's country, still, it is not to be understood to be necessary for it to halt in the midst of a series of successes for the purpose of making these establishments. Troops are left behind, or others are brought up, to occupy and fortify the points to be held, and establish dépôts of supplies, while the army continues to advance and profit by the successes already gained.

When the secondary base is not parallel to the original base, the more distant extremity must be specially strengthened by art, because it is more exposed to the attacks of the enemy. The other extremity, from its retired position, is more secure, but affords less protection to the army.

An oblique base gives an army the opportunity of threatening the enemy's communications and base, without exposing its own. Thus the army M (fig. 2),

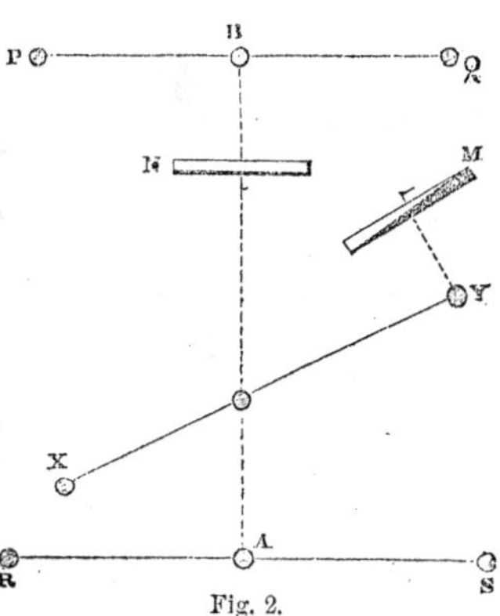

Fig. 2.

whose original base was R S, parallel to P Q, the base of the enemy, and which has as its natural line of operations A B, perpendicular to these two lines, could not make an attempt upon the left flank of the army N without running the risk of having its own communications intercepted, unless it had taken a secondary base, X Y, oblique to A B; for the positions would be reciprocal. If you cut the line of communications of the enemy, he may take possession of yours, which is left open to him. But with the new base, the lateral movement is not imprudent, because M always has a direct and safe retreat upon the point Y, which is near enough to give support.*

Attention has already been called to the advantages of a base of operations which is concave towards the enemy, and embraces with its wings the theatre of the war. When it is decidedly of this form, it gives an opportunity of changing the line of operations, if necessary, without the loss of the support of a secure base. Suppose, for example, that the base of operations has the form R S T (fig. 3); the army M, basing

* Points like the one in question, which are marked in the figures by small circles, are fortified cities or posts.

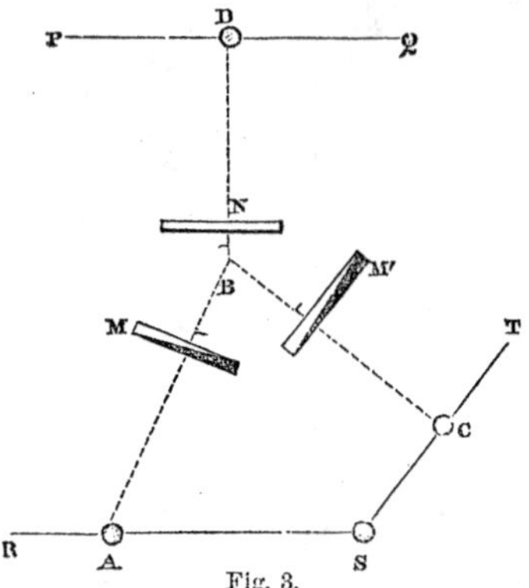

Fig. 3.

itself on R S, may at first adopt the line A B in its operations against the army N, whose base is P Q, and afterwards, if circumstances render it necessary or advantageous, the line B C may be taken up, which has S T for a base. In this way the King of Prussia, before the battle of Prague, passed from the left to the right bank of the Moldau. The passage having taken place, he had a new base upon the frontier of Silesia, and abandoned the one he had at first in the defiles separating Bohemia from Saxony. His first line of operations was from Dresden to Prague by the left bank of the Moldau; the second was directed towards Neiss, a strong position in Silesia; the two bases made an angle with its opening towards the enemy. In the same way, in 1809, Napoleon had the Danube for his base, and a line of operations directed towards Ingolstadt, but found it prudent to change both, previous to the combat of Landshut and the celebrated battle of Eckmuhl: he wrote to his generals that *he was no longer concerned about the Danube;* that he had chosen a new base on the Lech, and that they should fall back to Augsburg in the event of a failure. (See fig. 12, p. 161.)

When the change in the line of operations is made just before a battle, great advantages may be gained, because it deranges all the plans of the enemy. None but great generals, however, know when to make such a change.

The army N (fig. 3), which has but a single base, P Q, must remain upon the line B D, as it otherwise runs the risk of having its retreat cut off, in the event of a reverse. It is laid down as a principle, that in such a case, even if the forces were superior, the abandonment of the line B D, which is virtually the abandonment of the base P Q, would be wrong.

When one party controls the sea, a portion of the coast, where a good harbor may be held and furnished with supplies, is a sufficient base for an invasion. The English on different occasions have taken Corunna, Lisbon, and Dunkirk, as bases of operations; they used every possible exertion to retain the last-named place as a foothold upon the coast of France; they still hold Gibraltar, whence they are ready to penetrate into Spain whenever it is necessary or politic.*

Art. III.—Lines of Operations.

All routes are not equally good for moving against an enemy. Some lead more directly and safely than others to the object aimed at; some are more or less

* Vera Cruz was the base for the operation of General Scott's campaign in Mexico, in 1847, as also for that of the French, in 1862–'63.

favorable, according to the relative inferiority in cavalry or infantry; some abound more in resources, are more convenient for turning the enemy's positions; some are more completely protected by the base of operations, &c. One of the most valuable talents of a general is that which enables him to estimate properly the relative advantages of lines of operations, and to choose the best.

The line of operations is *simple* when the army proceeds in a single direction and remains united; or, at any rate, the corps composing it are not so distant from each other that they cannot afford mutual support. It is, therefore, the case of roads being used nearly parallel, not distant from each other, and separated by no great obstacles.

The line of operations is *double* when the army is divided, and, although proceeding from a single base, the different corps move along lines so distant from each other that they could not be concentrated the same day upon a common field of battle. When Marshal Wurmser, issuing from the passes of the Tyrol to succor Mantua, which was besieged by Bonaparte, divided his army into two parts, with a view of advancing simultaneously along the valleys of the Adige and Chiese, on both sides of Lake Garda, he had a double line of operations. In the same year, 1796, Jourdan and Moreau, basing themselves on the Rhine, between Basle and Mayence, manœuvred upon the Main and Danube separately: they had a double line of operations; or, what is nearly the same thing, each had his own line.

Except in the case of being as strong as the enemy, both morally and physically, or stronger upon each line, it is disadvantageous to use a double line of operations, especially if it is divergent; because the enemy may take a position between the fractions of the army, beat each separately, and find himself favorably situated for cutting their communications.

The greater the progress made upon divergent lines of operations, the greater does the disadvantage become; because the two armies are constantly increasing their distance from each other, as well as the difficulty of affording mutual assistance. The enemy between them may suddenly move against one of the two corps and beat it before the other can come up; he may then return to the latter, with a fair prospect of overwhelming it in the same way. In the first of the two examples cited above, Marshal Wurmser was beaten by his young opponent, who took a position between his two corps and destroyed them in succession. In the second example, the French generals were forced to retreat by the Archduke Charles, who manœuvred upon a single interior line, and had the skill to profit by his position. It is evident from this explanation, that although there is danger of being enveloped on a battle-field, the case changes when the distances increase, and the enveloping corps are too far separated to be able to combine their attacks.

Then the central position is the best, provided there is a display of the energy requisite in such circumstances. Thus, the rules of strategy may differ essentially from those of tactics; they may even be directly in opposition; and this is one of the causes which often render the application of these rules so difficult.

The only case in which, even with forces nearly equal, and when powerful motives urge to such a course, it is proper to take a double line in presence of a respectable enemy, is that where he moves upon two diverging or widely separated lines. But then the lines should be *interior*, in order that the two corps M and M′ (fig. 4) may, if necessary, keep up an inter-communication, give mutual assistance in case of attack, or concentrate suddenly against one of the hostile corps N N′, which are manœuvring upon *exterior* lines, and cannot possibly take part in the same action. The principle of interior lines, especially when they are converging, is really nothing but a modification of that of the simple line of operations. This principle consists in always keeping your corps between those of the enemy, more nearly together than his, and better

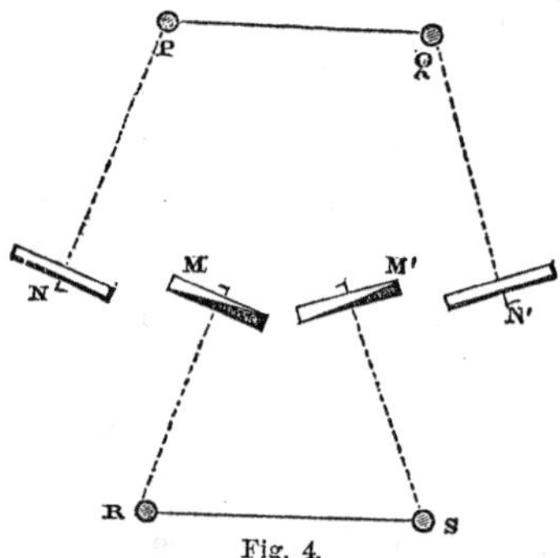

Fig. 4.

prepared to afford mutual support. It is, however, always safer to manœuvre upon a single line than upon two, even if they are interior; especially so when the country is entirely open, and there is no obstacle between you and the enemy which would permit him to attack one of your corps by surprise, before the other could rejoin it.

We must not confound with double or multiple lines of operations the different roads taken by the corps composing a single army, for the purpose of reaching a common place of concentration; in this case there is a common object to be attained, and the corps are separated only temporarily, to enable them to move more rapidly, to gain a thorough knowledge of the country, and to subsist more easily. This separation is, indeed, the perfection of art, when it is arranged in such a manner as to cease at the moment the decisive struggle begins. It is one of the best means of keeping the enemy a long time uncertain as to the point where he will be attacked. To know how alternately to separate the corps, in order to embrace a great extent of country when circumstances permit or require it, and to concentrate them rapidly to strike decisive blows, is one of the most evident marks of military genius. No general of modern times has given stronger proofs of possessing this faculty than Napoleon.

When a commander is forced by the events of the war to change the original line of operations, and to take a new one, the latter is called an *accidental* line of operations. But it would be improper to apply this term to a line taken voluntarily, for the purpose of marching upon a point that the enemy may have left unprotected, because he imagines himself threatened in another direction. The change of the line, far from being an accident in this case, is the result of a wise combination, and the probable cause of important successes. The primitive line would then be, to some extent, a pretended one, and the secondary line, the real one; it should not, therefore, be called accidental: it would be simply a new line of operations.*

It is sometimes proper, in a retrograde movement, to abandon the natural line of retreat and adopt a new one, for the purpose of drawing the enemy into a part of the country where his superiority may be diminished and himself removed from his principal object. The retreat is then made parallel to the frontier instead of along the original line, which is usually perpendicular to it. This line of retreat cannot be styled accidental, because it is made through choice and has an advantage over the natural line. It has received the name of *parallel retreat*, which is very appropriate. Frederick of Prussia, having been obliged, by the loss of

* Washington made a change in his line of operations after the action of Princeton, in New Jersey, which is one of the finest in military history. Napoleon had projected a change in his line of operations, in case he lost the battle of Austerlitz, but victory rendered its execution unnecessary.—*Halleck's Military Art and Science.*

a large convoy, to raise the siege of Olmutz, adopted a line of retreat through Bohemia in preference to re-entering Silesia by his natural line, from Olmutz to Neiss: by pursuing this course he continued, though retreating, to wage the war on his enemy's soil, and to relieve his own provinces from its burdens. The line was really accidental only for his adversary. Success, in such a case, requires, in the first place, that the force be not too much inferior to that of the pursuing enemy; and, secondly, that the retiring army be not so far from its own territory as to run the risk of being entirely cut off. The nature of the surrounding country must be considered in coming to a conclusion in this matter; if it is difficult, the movement will be the less dangerous; if the country is, on the contrary, free from obstacles, the safest road will usually be the shortest.

If the retreat is effected behind a line of defence which the enemy has forced, it may still be parallel rather than perpendicular to that line, because he is thus withdrawn from his object of penetrating into the interior and marching on the capital; there is thus a gain of time, and although the country is not relieved by it of the burdens of the war, it has all the other advantages of the parallel retreat. The retreat of Marshal Soult, in 1814, before the English army, may be classed under this head. Wellington's main object was to cross the Garonne and march upon Paris; Soult moved along the Pyrenees, forcing Wellington to follow him as far as Toulouse, where he detained the English general several days.

When a choice is to be made between several lines of operations, that should be preferred which offers the greatest facilities for subsistence—where the army will be safest and have the greatest advantages. If it is superior in cavalry, the level country will be better for its operations; in the contrary case, broken or mountainous country will probably be more suitable. A line of operations parallel and near to a water-course is very favorable, because the army has at once a support for one of its wings, and great additional facilities for transportation. The position of the enemy has also a great influence upon the choice of a line of operations. If he is spread out in cantonments over a large extent of country, that route will be the best which leads into the midst of his isolated corps, and gives an assailant the opportunity of separating them still more. If, in these circumstances, the line of operations were directed towards one of the extremities of the hostile army, its corps would be pressed back towards each other, and their concentration would be rather expedited than hindered. If, however, the enemy is concentrated, a direction should be chosen by which one of his flanks may be threatened, provided your own line runs no risk of being cut; the first principle is never to uncover the base or the line of operations. The plan to be adopted depends also upon the character and capacity of the opposing generals, the quality of the troops, their state of discipline, &c. Turenne, when he had to contend against the great Condé, by no means undertook what would have been simple and easy in presence of inferior men. Upon one occasion, in the campaign of 1654, he lost several men by passing before the Spanish lines within range, which gave rise to some remarks from several officers who were with him; he replied: "It is true, this movement would be imprudent if made in presence of Condé; but I desire to examine this position closely, and I know the customs of the Spanish service are such, that before the archduke is informed of our proceedings, has given notice of them to Prince de Condé and received his advice, I shall be back to my camp." There spoke a man who was capable of judging in those matters which belong to the divine part of the art. It is in these delicate distinctions that is manifested a true genius for war.

It has been stated that the line of operations should be directed upon the flank of the enemy when he keeps united, if it can be done without exposing your own. It almost always happens that the *turner is turned;* and this is strictly true in a country free from obstacles, when the bases of the two armies are nearly parallel and equal. In fact, the army M (fig. 5) cannot make an attempt upon the communications of the army N, except by following a line of operations, S B, which is oblique to the base, R S,

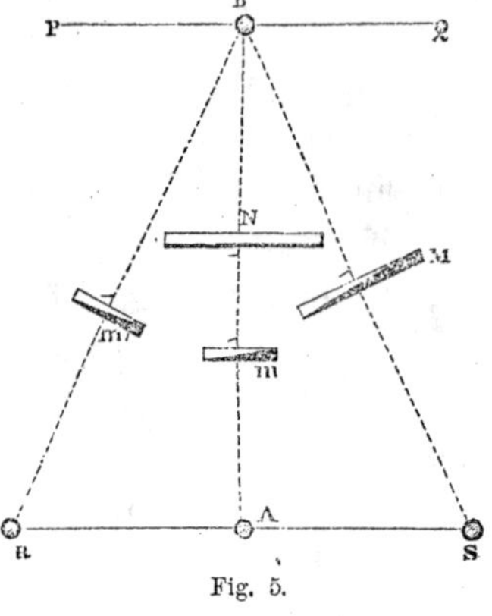

Fig. 5.

and supported by it at its extremity; but then the central line A B and the communication on the left are uncovered, and the advantages of the extended base are lost, since the only line of retreat left is upon the point S. It is then only in the case where natural obstacles enable a few troops to defend the communications, B R and B A, against attack from the enemy, that the principal line of operations may probably be directed towards the flank of the enemy, because by throwing forward the small corps, *m* and *m'*, upon the other lines, the base, R S, is covered, while at the same time the enemy is threatened upon several points. It may be stated generally, that it is only when favored by natural obstacles in this manner, movements may, without very great risk, be made against the flanks of an enemy. The influence of the features of ground is always manifested in the resolutions arrived at in war; the study of topography is, therefore, of the highest importance for officers of all grades, but especially for those who are intrusted with important commands.

A line of operations which is too extended, relatively to its base, loses a portion of the advantages it may possess, on account of the opportunity afforded an enemy of cutting it. It is this consideration which induces a general, as he advances from his primitive base, to adopt secondary ones. To desire to establish a fixed relation between the length of a line of operations and the extent of a base would be ridiculous pedantry. The triangle formed by connecting the position of the army with the extremities of the base may vary infinitely. Its form and dimensions generally do not depend upon our will. It is sufficient to say, that the greater the extent of the base, the more distant may be the operations of the army without having its communications endangered; nearly in the same way that a pyramid may, with equal stability, have a greater height as its base is broader.

Art. IV.—Strategic Points.

Objective points are also called *strategic points;* and we include under this denomination not only those which may be considered as the principal objects to be attained, but also those whose occupation is of manifest advantage to the army.

A capital city is a strategic point of great importance, because it regulates or greatly influences the public opinion of the nation, contains abundant resources of every kind, the loss of which may greatly paralyze the enemy; and to the minds of a large portion of the people all hope of successful defence is lost when an invading army has reached the heart of the state. If by seizing the capital no other than the moral effect is produced of discouraging the people, it is still a worthy object of the efforts of an invading army.*

A point is strategic when it is the key of several important communications, when it controls the navigation of a river or defends some important passage. For instance, Ratisbon, on the Danube, which is the centre of important communications on both banks of the river, has always been shown, in the wars in Germany, to be a strategic point for the possession of which the belligerents have made great efforts. Smolensk on the Borysthenes is another; this city is in the interval between the Borysthenes and the Méja, the key of the Russian empire on the side of Moscow. Hence, in 1812, the Russians were as anxious to hold this point as the French to seize it. The city of Langres, situated near the sources of several of the

* The possession of Genoa, Turin, Alexandria, Milan, &c., in 1796, both from their political and military importance, had a decided influence upon the results of the war in these several states. In the same way Venice, Rome, and Naples, in 1797; Vienna, in 1805 and 1809; Berlin, in 1806; Madrid, in 1808; and Paris, in 1814 and 1815. If Hannibal had captured the capital immediately after the battle of Cannæ, the Roman power would have been destroyed. The taking of Washington, in 1814, had little or no influence on the war, for the place was then of no importance in itself and was a merely nominal capital. It, however, greatly influenced our reputation abroad, and required many brilliant successes to wash the blot from our national escutcheon. The possession of the capital is of prime importance in a civil war, as recent events in America have proved.—HALLECK.

streams which water France, is another strategic point. Alexandria, in Piedmont, as a fortress beyond the Alps, is equally so; its possession is indispensable for holding the plains of the Po. Soria, upon the plateau of Old Castile, is also a strategic point, although that city has nothing in itself to make its conquest desirable; it is rather the position which is important, because, like Langres, it is near the source of several rivers.* In level countries there are few points that are strategic from their positions; they become so on account of the fortifications which surround them, as well as the supplies and resources they furnish. Many of these points are found in wooded, undulating countries, cut up by rivers and lakes. Such countries are usually thickly settled.

In mountainous regions, like the Alps and the Pyrenees, the number of strategic points is somewhat limited, but they are very distinctly marked: they are found at the entrances of defiles, at the points whence several valleys branch out, at the junctions of roads. The high ground at the point where several ridges meet is also strategic, for the troops occupying it may select the route by which to descend: their movements are from a central point outwards, while those of the enemy can only be accomplished

* Cairo, at the junction of the Ohio and Mississippi Rivers, is a strategic point of great importance. Paducah, at the mouth of the Tennessee, is also important. New York, New Orleans, Fortress Monroe, St. Louis, and Chattanooga are strategic points.

by long circuits made to pass around the mountains which it may be impossible to cross. If he has once fairly started in a valley, he is in a measure forced to remain in it, as it is extremely difficult for him to pass from one to another; but the troops stationed at the head of the valleys may use them at pleasure and in turn if necessary.

Art. V.—Plan of a Campaign.

Before undertaking a war or any operation, a certain object to be attained must be decided upon; the direction of the necessary movement must be fixed in advance and as accurately as possible; there must be a thorough understanding of the real object aimed at, in order that every thing may be prearranged and nothing left to chance. This is called *making the plan of the campaign:* the definition itself implies that the war is to be offensive.

It is evident that this plan should be limited to the grand strategic operations and be only an outline of them, leaving as much latitude as possible for choice of movements of detail and execution. It would be absurd to pretend to dictate to a general what he ought to do from day to day, for, after the first marches, and he has reached the theatre of the enemy's operations, he no longer does what he likes but what he can: marches, manœuvres, combats depend upon imperious circumstances; the determinations to make them are usually sudden and result from the position, resources, forces, and condition of the enemy. The general should be perfectly untrammelled in carrying out a plan of campaign, even if it has been marked out for him by higher authority; but it is still better, after selecting a competent man, to let him make his own plan and attend also to its execution, as he, being more interested in its success than any one else, will strive to make it succeed.

In making a plan of campaign, maps are necessary, and for this purpose those upon a small scale are the best, as they show a whole country together. It is enough for them to indicate exactly the position of fortified places, water-courses, the main roads, mountain ranges, frontiers. Those maps are best which are most distinctly marked, because they are more easily consulted. Maps on a large scale, or topographical maps, are useful in determining questions of detail, but the number of sheets such maps usually consist of renders them inconvenient for use in making the general arrangements of a plan of a campaign. The topographical maps will be consulted when a camp is to be established, a position taken up, an order of battle agreed upon.

The plan of the campaign indicates the places of concentration of troops, the base and line of operations, the strategic points to be attained.

The choice of the points of concentration of troops is regulated not only by the consideration of subsistence, although this is always very important, but by the desire to occupy such positions as will keep the enemy uncertain where he is to be attacked, and which permit the initiative to be taken promptly and in strong force. The distance of the different points of assembly should be so calculated that all the corps may arrive at the same time at the general rendezvous in order to pass the frontier together. Thus, in 1815, Napoleon having determined to attack the allied armies, directed his forces towards Belgium, which was occupied by the English and Prussians. They were covering all the space between Liege, Mons, and Brussels, for a distance of sixty or seventy miles along the Sambre and the Meuse, being principally established between Mons and Namur. Napoleon assembled his forces at Maubeuge, Beaumont, and Philippeville, thus threatening Mons, Brussels, and Namur, and forcing the enemy to remain separated. He resolved to pass the frontier at Charleroi, with a view of assailing the centre of the hostile cantonments. Calculating the distance, he caused the several corps to set out at hours which enabled them to arrive simultaneously at Charleroi. The place was readily taken. They passed the Sambre, and on the same day, June 15th, camped in the enemy's country. This is an interesting example: it demonstrates the influence of a good choice of places of assembly upon the first successes, the moral effect of which is so great. The same general, two years previously, operating upon a vastly greater scale, obtained the same precision in the results of his arrangements. He had assembled his corps on the Niemen; he sent them by different roads and at different times towards Ostrowno, where they arrived, after long marches, at the moment when the two armies were about to come to blows. General Barclay, with Fabian prudence, avoided the effect of this formidable concentration, by abandoning the position to his adversary.

In a plan of campaign, the base of operations can only be a subject of discussion when there is a choice between two frontiers, and there is a hesitation which to take. For example, if a war were impending between France and Austria, and France, as she usually does, wished to take the offensive, there would be occasion for discussing the relative advantages of the Rhine frontier or that of the Alps. This is the time when the general form of the base must be considered, and its influence upon the object in view. If it is concave, the army after advancing will have a secure support for its wings and rear. If it forms a salient angle, as does Silesia in the Austrian States, there is the advantage of being enabled, with a single body of troops, to keep the enemy uncertain as to the quarter to be attacked; and he is thus obliged to spread his forces along the whole circuit to be guarded, which is very unfortunate for him. But even if he remains united, there is every facility for attracting his attention in one direction, suddenly striking a blow in the other by the shortest route, and taking him unawares. This salient form is therefore advantageous for opening a campaign; but the army runs the risk, in case of a defeat, of being separated from its base. The case is reversed with a concave base.

The importance of the base of operations increases with its extent, its defensive properties, its resources of every kind, its vicinity to dépôts, the facilities for reaching it from the interior, &c. If it is perpendicular to the proposed line of operations, and extends beyond it on both sides, it is better than if it were oblique, and only supported the line on one side.

When, therefore, in a plan of campaign, there is an opportunity to compare the relative advantages of two bases of operations, not only should the character of the base be considered, as a more or less solid support for the army, a greater or less obstacle to the enemy, but also its form and relative position.

Of two frontiers, one necessarily has preponderating advantages; it should therefore be selected as the base of operations, and all the disposable means of attack should be collected upon it. A fault is committed in proceeding from two bases at the same time. It is much better to stand on the defensive on one frontier, in order to collect more troops on the other, and secure more chances of success. Success seldom follows a division of forces; and the same reasons which make double lines of operations dangerous, are equally powerful against attempts to act offensively from two distinct bases. There should, on the contrary, always be the greatest possible concentration of effort, in order that successes which are gained may be decisive; and they will always be more so upon the important frontier than upon the other; all proper means of strengthening the one, without too much weakening the power of defence of the other, should be adopted; in the same manner upon a battle-field, troops are withdrawn from the wing that is less threatened, in order to re-enforce that one which is to decide the fate of the conflict.

The discussion of the different lines of operations that may be followed, is an essential part of the plan of a campaign. In fact, it is not with these lines as with the base, which oftentimes gives no ground for discussion, for the reason that it is determined beforehand by the geographical position of the country where the war is to be waged. Lines of operations, on the contrary, may be very different, because there are almost always several routes leading to the same object. There are so many things to be considered, that the choice of the best line is always a very delicate matter. In this, the general shows his true conception of the principles of the art. It is scarcely possible to lay down other rules on this subject than those stated in Article III. It may be added that, in discussing these lines, effective distances are to be calculated, that is to say, the times actually necessary for passing over them, and not the distances on the map; all things else being equal, the shortest route should be adopted; the best road will be used when there are no strong reasons for leaving it, for troops march much more rapidly along a broad and convenient road, than through narrow and difficult paths. The character of the cities to be passed, defiles, obstacles, may oblige the use of indirect and bad roads; for the mere difficulties of a march are not usually so great as those of capturing cities and forcing defiles. When the army, in advancing, is enabled to rest one of its wings upon a natural obstacle, the best line of operations will be the one which is nearest to that obstacle, because during the march it will be better covered by the army, whose front is almost always several miles in extent; at any rate, while a battle is not imminent. When the two wings of the army are unsupported, the line of operations should proceed from the middle of the front, in order to be, as far as possible in this unfavorable case, equally covered on both sides. As a general rule, in a march as in a battle, the line of operations must not be uncovered; all the arrangements should, on the contrary, tend to cover and defend it as much as possible.

The choice of objective points also requires much sagacity, as the possession of one may give great military advantages, such as the control of the navigation of a river or the passage across it, or the control of several roads or valleys having a common point, or an essential line of retreat, &c.; another may have advantages of a different kind, but no less important, such as provisions, munitions, arms, clothing, equipments, tools, money,—all indispensable articles, which must be carried with the army, if they cannot be obtained along the way. The capital, large cities abounding in resources of every kind, are the principal points to be occupied, and are those which, in a plan of campaign, should figure as objective points of the first importance. But this subject has been already treated to some extent in the preceding article.

Art. VI.—Plan of Defence.

The plan of a defensive campaign may be better termed *plan of defence.* It consists, in the first place, in determining the character of the war to be undertaken, which depends on national character, the resources of the country, its topography, and climate. The French defend themselves by attacking; the Germans patiently wage a methodical war behind their frontiers; the Spaniards strive to exterminate their adversaries in detail; the Russians, in 1812, were seen to devastate their country and burn their capital, to deprive the invader of necessary resources. People who are courageous, but unaccustomed to privations, will not protract a war, but strive to end it by brilliant actions; they will strike heavy blows, preferring a single great battle, where defeat even will be glorious, to a series of partial engagements, which exhaust the country and lead to no decisive result.

To carry the war into the territory of the invader or of his allies, is a sure means of throwing part of the burden upon their shoulders and thwarting their plans. The *morale* of troops is raised by a proceeding of this kind, and the chances of success are increased. But to reap success from such a plan, the forces must not be too inferior, and the conformation of the frontiers must be favorable. The army must not go too far from its own territory, in order to avoid being compromised. The zone of country between the army and its frontiers will supply its wants, and should be defended to the last extremity, by bold attacks, with the entire disposable force, upon that corps of the enemy which is most conveniently situated for the purpose. If the stroke is successful, the army may pass to the offensive, giving an entirely new phase to the war. If it is unsuccessful, there should be still greater concentration, strong positions for camps should be selected, or the army should retire under shelter of some fortified city, or cover itself by means of a river, &c. You thus oblige the enemy to follow you, and to move in a zone of which the devastation is entirely at his expense. Montecuculi, in his Memoirs, strongly recommends this kind of a war for the defensive. He says: "On the territory of an adversary, malcontents are enlisted in your favor; there is no need for concern about supplies of men, money, and other necessaries, as these only become exhausted in the theatre of war, which is upon your adversary's soil."

An army fighting upon its own territory causes,

necessarily, much injury to the people, but there are less risks to be run; the population is favorable, and gives assistance of every kind; every individual is a spy upon the enemy, and gives information of his operations; the battles are fought on positions examined, and sometimes prepared, in advance, where the enemy is obliged to come after the army; it may move in any direction, because every space not occupied by the enemy may serve for a base of operations; there is every facility for threatening his communications; he is forced to make detachments for the purpose of guarding the cities he has taken, and to secure his subsequent movements. These detachments are fair game—they may be separately attacked and beaten, or surrounded. When fortune crowns with success the efforts of the defenders, their victories are much more decisive if obtained within their own territories than beyond them, because the vanquished enemy has defiles to repass, and a retreat to effect, in the midst of a population which is hostile and ready to fall upon him. But, in order that such a method of acting defensively be successful, there must be unity of intention and effort in the nation,—no civil discords to divide it into two hostile parties,—and the dread of foreign domination must exist in every heart. What can a divided people do in presence of an enterprising enemy? What can the army do, even when most faithful to its flag, its own honor, and organized in the most perfect manner, if it is not seconded by the population,—if it meets nothing but inertness and lukewarmness, or hostility, where it should receive assistance, support, co-operation? This essential condition, without which defence is impossible, should, therefore, be assumed as satisfied in treating a question of this kind.

A frontier which has its convexity turned to the enemy is favorable for a defence made within its limits at the opening of the campaign, and may lead to the adoption of that method. In fact, from a central point which is strongly held, the whole circuit of vulnerable frontier may be observed, and the army may stand ready to move by the shortest route to the point threatened. Thus, the Piedmontese, by occupying a position near Turin, may reply to any attack made upon them from the surrounding Alps.

The character of the war being decided upon, the details are next to be considered, and these properly compose the *plan of defence*. The determination of the places where a suitable resistance may be made; of those to which the forces shall retreat in case of reverse, and the routes to be pursued; the disposition of the troops while expecting the enemy's approach; the manner of anticipating him, wherever he may come; the indication of points for concentration behind the frontier, as soon as his projects are developed; the method of supporting the advanced corps, and those in observation, by central reserves; finally, the designation of points to be fortified by the appliances of art, bridges to be destroyed, roads to be mended, &c.: these are the objects to be attended to in making a plan of defence.

The local topography will indicate the appropriate military dispositions to be adopted. It is scarcely possible to lay down rules on this point; all that can be said is, that a too great scattering of the troops is always dangerous; consequently, far from thinking of guarding all the passages, some must be abandoned, that others, the most important and the most threatened, may be better defended. If the enemy leaves these to go to the former, his projects must be counteracted by similar means, an effort being always made to oppose to his advance, by whatever route he may come, as many troops as possible. Hence we see how greatly the defence of a State is influenced by the form of the frontiers, the direction and character of the roads. If you are enabled to follow, by straight interior lines, the movements of the enemy manœuvring on the circumference of the circle, you have all the advantages of position, and may always anticipate him at any threatened point. Even in the contrary case, you cannot possibly close all the avenues; that would oblige you to form a long cordon of troops, very weak at every point, through which an enemy may penetrate wherever he pleases. You should rather place the main body of the troops behind the frontier, in a well-selected position for the promptest movement upon the several roads open to the enemy. In advance of this corps, and upon its flanks, you should send out simple detachments to occupy the passes temporarily, and give notice of an attack. They will dispute the ground while falling back, thus retarding the march of the enemy, and giving time for troops to approach and meet him. In this way all the surrounding country will be efficiently watched to avoid surprises, and the troops are, also, readily concentrated to fight. Such are the general arrangements to be adopted in a plan of defence; evidently great latitude is given them, and they should conform to the requirements of localities. If, however, they are based upon the principle of *concentration*, they will be always preferable to an insecure cordon, whose detached bodies afford no reciprocal support, are withdrawn from the immediate command of the general, who cannot be omnipresent, and, on account of their distance apart, they cannot be concentrated when the line is broken anywhere. Feuquières blames Marshal Catinat for neglecting these principles: "Catinat being intrusted with the defence of the passes of the Alps against Monsieur de Savoy, adopted a cordon system, scattering his forces along the whole circuit of the mountains, and thus gave the enemy an opportunity of taking the offensive, although much inferior in infantry, which is the important arm for mountain warfare. De Savoy, by concentrating his forces, was superior to Catinat wherever he chose to attack, especially as he occupied the centre of the circle and could threaten several points at once, and choose for the real attack that which seemed best."

To give prompt notice of the enemy's movements a system of signals must be used, or the magnetic telegraph.

It is not always by taking position in the direct path of an enemy that his advance is opposed, but sometimes points may be occupied on the flank with much advantage, so as to threaten his line of operations if he ventures to pass. If these flank positions can be held in force, the enemy must leave his direct route to attack them, for, unless he carries them, he runs the risk of being cut off from his base. He is thus compelled to fight on ground selected by his adversary, and prepared for his reception by fortification and any other means circumstances may permit. In this way the Turks, by concentrating at Shumla, completely arrested the Russian invasion of 1810, and by pursuing the same course in the war of 1828, they prolonged their defence during a whole campaign. If, in 1812, Kutosow, instead of covering Moscow and taking position upon the plateau of Mojaisch, had retired towards Kiow, he would, as Napoleon himself admits, have drawn the French army in that direction, and the immense sacrifice of Moscow would have been avoided.

Next come the successive lines of defence presented by a country, which are naturally indicated by the water-courses and the chains of mountains or hills. The advantages and disadvantages of these are to be carefully discussed. The best lines are those whose wings are supported by natural obstacles, such as the sea, a great lake, a broad and rapid river; which have a controlling influence over the surrounding country; are only passable by a small number of roads easily guarded; whose general form is convex to the front, and behind are good roads facilitating rapid movements, and by short marches, to any point attacked. The famous lines of Torres Vedras, constructed by Wellington, in 1809, to cover Lisbon, satisfied the most of these conditions; they formed a great curve, several leagues long, one of its extremities resting on the Tagus and the other on the sea; there was a chain of hills, most of them crowned by fortifications, behind which the English army could manœuvre unseen, and move to different points of the curve by roads which had been put in good condition, and some newly constructed. The French could not force these lines, although they were not covered by a river in front of them. Continuous artificial lines, erected by the labor of men, deserve but little confidence, on account of the difficulty, not to say impossibility, of defending these inert masses when unprotected, or but feebly defended by natural obstacles. The lines of Wissembourg, although covered by the Lauter, did not prevent the French from being driven out of them by the Austrians, under Marshal

Wurmser. Resting at the left upon the Vosges mountains, and at the right upon the Rhine, they were ten or twelve miles long; it was impossible for an army of less than 20,000 men to make an energetic resistance in them. But a few hours were necessary to see them entered at several points, and the defenders driven off. The most celebrated artificial lines are the Chinese walls, and these never arrested the invasions of the Tartars.

The advantages to be drawn from fortified cities should be considered, either for protecting against the attempts of the enemy all the supplies that must be collected for the troops, or for the defence of certain points which should be held as long as possible. A city surrounded by a simple wall and ditch, which secures it from danger of being taken by surprise or open assault, may, if favorably located, be of the greatest value. A city which will withstand attack but twenty-four hours, may be very useful, as in this space of time events may transpire that will entirely change the appearance of affairs, and save a State from the greatest of all misfortunes—the loss of its independence. If, in 1814, the city of Soissons had closed its gates and made but a semblance of resistance, Blucher could not have escaped Napoleon's active pursuit, and perhaps France might have been saved. In 1809, the Austrian Government rebuilt the walls of Comorn, which had been destroyed thirty years before, through Joseph's want of foresight.

In a special manner cities upon rivers may be very valuable to the defence, although they may not be in a condition to repel an attack in force. By giving possession of the bridges, they facilitate manœuvres from one bank to another, enable an army to cover itself by the river or to pass it, in the event either of its being too weak to march against the enemy, or of a good opportunity being presented to take advantage of some false movement upon his part. If the army passes the river, these cities are so many *têtes-de-pont* securing its retreat and covering its operations after repassing; they are so many points by which the army may debouche at any instant, and the enemy is thus kept on the watch and obliged to divide his forces. In a word, fortified cities give much value to the lines of defence formed by rivers; they may even make a river, which is perpendicular to the frontier, a dangerous object for an invading army, by the facilities afforded the defenders for occupying the side opposite the enemy and threatening his communications. He must then halt and gain possession of the cities which control the river. But this is a very difficult operation in presence of an adversary who is determined to make a vigorous defence. If the attacking party advance along both banks, his army is divided. The defensive party, holding the bridges, may assemble in force and operate against either subdivision of the hostile army. His success is altogether probable. The invader must therefore remain united and turn his attention to the river whose *têtes-de-pont* are so threatening to him, just as would be necessary for any flank position strongly occupied, that could not be passed by without great danger. If such a river had upon it no fortified places, it might, instead of being an obstacle, form a support for one of the wings of the invading army, and facilitate its transportation.

It is therefore true that cities commanding the water-courses are of great importance, and deserve careful attention upon the part of those engaged in preparing a plan of defence; even when they are imperfectly fortified, their advantages should not be neglected, but, on the contrary, efforts should be made to increase their powers of defence.

Positions, properly so called, that is, the places where an army may advantageously fight, should be indicated in a plan of defence. The relative advantages and disadvantages of the several positions should be carefully examined, in order that the former may be profited by, and the latter avoided as far as the circumstances permit. Calculation should be made of the efforts necessary to augment the defensive power of the several positions by the means presented by the engineer's art, and to remove the obstacles offered to the free movement of troops; an enumeration should be made of the troops of each arm of the service necessary or useful in the defence, and the places for them to be camped or quartered should be indicated, as also the resources in the neighborhood of each. As a measure of prudence, in anticipation of a reverse, the question of the best line of retreat for the army should be discussed; what are the best places for rallying; what parts of the line are to be disputed foot by foot, and what parts are to be rapidly passed over, &c. The conditions of a good military position will be mentioned under the subject of battles.

The roads are to be designated that should be broken up to make them impassable by the enemy; also those to be improved, in order to facilitate the defence. Too much attention cannot be bestowed upon these objects, which at first sight seem of secondary importance, but have so much influence upon the rate of movement of the two parties, and consequently upon the execution of any project. A well-organized government will always take care, in the establishment of canals and roads, that the advantages procured by them to commerce are not very dearly paid for in war. The best road may be secured from danger by constructing a work for its defence at the most suitable point. Roads may undoubtedly, and especially in mountains, be cut and rendered temporarily impassable, by the destruction of bridges, &c.; but, independently of the greater or less facility with which an enemy may repair these injuries, there is always a feeling of repugnance to taking steps for the destruction of the communications of the country; it is delayed to the last moment, and often until too late. The safest plan is to build the forts, of which mention has been made, and for this reason a plan of defence should indicate the positions for them. While marking the points where a defence may be made, the method of obstructing the roads in front should be indicated, as well as the means of improving the parts to the rear, with the double view of retarding the approach of the enemy, and facilitating the arrival of succor.

All the roads connecting two positions, or which, being behind a line of defence, run parallel to it, should be repaired; for these roads are the routes to be passed over to succor threatened points, or to fall in force upon detachments of the enemy. The plan of defence should treat these operations in detail.

These are the essential objects to be mentioned in a plan of defence. There are others, no less important, which relate to the administrative department, such as hospitals, equipment, accommodating troops at the points of concentration, their subsistence, foraging, the postal service and other means of correspondence, &c. These objects should all be considered in a plan of defence, but it is impossible to go into all their details here.

Art. VII.—Strategical Operations.

Preparatory movements, marches skilfully made, with a view of taking position upon vulnerable points of the enemy's line; in a word, strategical operations, procure those immense results which sometimes follow a single victory. A battle gained is an excellent thing, but its consequences will vary according as the victor is so situated as to be able to profit by it, and cut the enemy's communications, separate him from his base, disperse his army, &c., or is unable to interrupt his retreat. In the first case, the victory is decisive, if the pursuit of the defeated enemy is vigorously prosecuted; in the second, he may be expected soon to make another stand, and fight again. When Napoleon moved his army from Boulogne, in 1805, by forced marches, to attack the Austrian army, that had already invaded Bavaria, and was advancing between the Danube and Lake Constance, he directed the different corps upon the flank and rear of the enemy, assuming as a base the Rhine below Strasbourg and the Main, of which he was master; he turned the mountains of the Black Forest, in whose defiles he showed several heads of columns, to cause Mack to think he was going to meet him by the way of the valley of the Danube. When Mack perceived his error, it was too late; the blow had been struck. He was cut off from his communications, beaten in several encounters, and obliged to take shelter in Ulm, where he capitulated with his whole army. The wise and rapid marches which preceded the combats of Gunzbourg, Elchingen, &c., were the causes of the successes of Napoleon, much more than those combats, brilliant as they were. The Austrians were completely cut off from retreat; they were enveloped, and forced to lay down their arms before the arrival of the succor they expected.* Marshal Saxe was convinced that marches contributed to the successful issue of a campaign more than battles, and he laid down the axiom, that "success in war is due to the legs of the soldiers." History justifies his assertion.

We shall now lay down the principles relating to marches, or rather to strategic movements.

Forward Movements.—A numerous army is always divided into several corps, which move along different roads, either to procure subsistence more easily, or to facilitate their deployments and preparations for battle. These columns should be the nearer to each other, as the enemy's enterprises are the more to be apprehended. If he may attack, the columns should be near enough to afford mutual support; there should be no obstacle between them, which will prevent their junction and concentration on the field of battle. But there is no necessity for the columns to move exactly side by side, and along parallel roads which have been opened with the axe and the pick. Such a course would make rapid movements impossible, and would give all the advantages to the antagonist who was not so precise. Since the time of Turenne, Luxembourg, and Villars, rapid marches had been forgotten; a weak army was divided up into numerous columns, which, with great labor, made roads through

* For the details of these marches, see Chapter III., Article III.

forests and across ravines, without losing sight of each other, so to speak. The result was great confusion, and inconceivable slowness. The French Revolution put an end to that folly, and a return was made to the free, rapid, bold marches of the Romans. So long as the manœuvre is out of range of the enemy's cannon, the intervals between the columns executing preparatory movements may be greater or less, according to circumstances; the only limit to be laid down is to keep the intervals such that the corps may be able to concentrate the same day upon the same field of battle.

Each column should take proper precautions to prevent being surprised, and its march should be preceded by an advanced guard. Negligence in this particular may lead to disaster, as history proves conclusively. The defeat of Flaminius at Thrasymene (now Perugia), is a celebrated example in point. This general imprudently ventured into the defile between the lake and the mountains without sending out an advanced guard, and without examination of the heights. He was hurrying to attack Hannibal, who was laying the country under contribution; his haste was so great that he would not wait for his colleague, who was coming up from Rimini with an army. In the midst of the defile the wily Carthaginian had taken a stand to bar the way. The action begins, and the Romans see descending from the surrounding heights numerous troops, to take them in flank; at the same time the cavalry, which they had passed by unseen, attacks them in the rear. The Romans, being obliged to fight on all sides, are obliged to yield, and the Carthaginians made terrible slaughter among them.

To make a march safe against the attacks of the enemy, it should, as far as possible, and in the manner previously explained, be along a river which will cover the flank of the column; but, even in this case, the avenues must always be occupied by which the enemy might approach. For this purpose troops are detached to take position and cover the march until the column has passed so far that there is no longer danger. The detachment will then rejoin the army as soon as practicable.

Those marches which are concealed from the enemy lead to the most important results. By such marches a general succeeds in placing his army upon the flank of his adversary, threatens his base, or surprises him in his cantonments. The most difficult countries are the most favorable for concealed marches, as there are greater facilities for hiding them; and the enemy, trusting to the natural obstacles, neglects ordinary precautions, and omits to seek for information. With patience, labor, and preseverance, material obstacles may be always overcome by an army that is not disturbed by the presence of an enemy. It may be stated that in this particular nothing is impossible. As an example, take the extraordinary march of Hannibal across the marsh of Clusium. He had two roads by which to advance on Rome; one through the defiles of the Apennines, easy but long, and held by the Roman army; the other shorter, but across vast marshes, supposed to be impassable. Having ascertained the depth of the marshes, and being convinced that the difficulties they presented, though great, were not insurmountable, Hannibal chose that road. Knowing he was not expected in that direction, he saw that he would have to overcome only local difficulties, and he was sure of success as far as they were concerned. By taking that road he avoided the defiles, where his cavalry, which was superior to that of the Romans, would have been useless, and where the individual valor of the Roman soldiers would have compensated for the incapacity of their consul. He had, moreover, good grounds for hoping that, having once effected the passage, he could inveigle the imprudent Flaminius into a battle before the arrival of his colleague, as in fact happened. The success of that bold movement was perfect, because it was so secret and prompt that the Romans had no time to offer any opposition.

Rapidity is one of the first conditions of success in marches, of strategical movements, as of simple manœuvres. By celerity of movement a general preserves the advantages gained by a fortunate initiative, and he follows up and completes the success a victory has begun. It is only by rapid marching that *war can be made to support war.* By remaining but a short time in one place, an army does not exhaust the provinces through which it passes; their resources, uncertain as they may be, will always be sufficient for the temporary wants of armies. By adopting such a system, there is no longer a need of the immense wagon-trains used to transport provisions; nothing is carried but what is indispensably necessary; the soldiers may be required to take a few rations of bread, and are followed by droves of animals to furnish meat. The troops, being thus unencumbered, can undertake and execute the grandest projects. It is impossible to gain any success in war without rapid marches. Proceed at a snail's pace, and you will accomplish nothing; misfortune will attend you continually, and the elements will seem to conspire against you.

If the enemy approaches, spreading out his corps over a large space, with the intention of enveloping you, move at once against him, and strike at the centre of his line. Endeavor to attack one of the isolated parts of his army, and defeat it before the others can arrive; move rapidly against another, and treat it in the same way: you will thus oblige them all to retreat, to take divergent roads, and to encounter a thousand difficulties in effecting a junction, if indeed it be possible for them to do so at all.

In this way Napoleon, in the beginning of his career, rushed from the maritime Alps, and fell like a mountain torrent upon the Austro-Sardinian army, commanded by Beaulieu and Colli, who had committed the fault of spreading out their forces in a long line, with the expectation of enveloping their young adversary. Colli was first beaten at several points; Beaulieu's turn came next, and from that time the two armies were so entirely separated, that one of the generals, desiring to cover Turin, retreated towards that city, whilst the other fell back upon Milan without attempting to rejoin his colleague. Colli was forced to accept such terms as the victor chose to give; Beaulieu could only stop his retrograde movement under the walls of Mantua.

But if the hostile army is concentrated, an attempt may be made to advance upon him along two lines towards his flanks, which will induce him to divide his forces in order to meet this double attack. Then the two corps, which have been separated only temporarily, with a view of immediately effecting a reunion, will move towards each other, and make a combined attack against the nearest of the two portions of the opposing army. It readily appears that such movements may be attempted only when the topography of the country favors, and to some extent suggests them. For example: a general, having his army covered by a river, need have no alarm at seeing his adversary appear opposite in the interval between his two corps, when he can make a rapid flank movement by means of roads along the river. In this case he may, with no great danger, pass to the right and left, in order to cause uneasiness to the enemy at two distant points. If the enemy weakens his centre to re-enforce the wings, the general may make a rapid and concealed night march, hurry back and pass the river, using temporary bridges for the purpose, for which the materials should all have been prepared in advance, and as secretly as possible. The river once crossed, he has the interior position; activity and bravery must accomplish what remains to prevent the separated corps of the enemy from effecting a junction, and they are thus forced to operate on exterior lines. Here we see the importance of making a good use of time, in order to concentrate, pass the river, and beat in succession the fractions of the army between which we have fortunately succeeded in interposing our own. Time is all-important in war; often an hour lost cannot be regained; in one hour we may be anticipated upon the decisive point; in one hour we may be overwhelmed by superior forces suddenly assembled, or we may let slip advantages which fortune grants only to activity and boldness.

The preceding remarks prove that if a frontier is attacked on several points, the defensive army should not be divided into the same number of equal corps, because the resistance will be feeble at every point; but the different attacks should be met by small corps, placed in observation chiefly, while the principal mass is kept at some convenient point, in order to fall upon a single one of the attacking armies when separated from the others. Suppose we take the case of 80,000 men being called upon to resist 120,000, divided into three equal masses of 40,000 each; if the defensive army is divided into three equal corps, each will contain about 26,000 men, and consequently there will be an inferiority of force at every point. If, instead of that arrangement, each of the advancing armies is met by a simple corps of observation of 12,000 to 15,000 men, they will be able to delay the enemy's march, and there will remain a central mass of 35,000 to 40,000 men, which, being united to one of the corps of observation, will form an army of about 50,000 men, and this may defeat the enemy, all other things being equal.

When the disparity of force is still greater, every effort must be exerted to be superior in numbers upon some point; but if that cannot be done, and it is sometimes impracticable, the rule must still be followed of concentrating as large a force as possible to meet the enemy. Then the struggle is to be prolonged by the development of the greatest attainable vigor and activity, and by skilfully using all the topographical advantages of the country. Napoleon, in the war in France, gave a striking example of the effectiveness of central manœuvres; he fought triple his forces for four months. He moved his reserves from point to point with marvellous rapidity. One day he fought a battle in one place, and the next he was twenty-five or thirty miles distant, marching to attack another enemy, whom he astonished by the exhibition of so much resolution and celerity of movement. Cæsar, when surrounded in the middle of Gaul by nations in insurrection, extricated himself from his critical position by similar movements. He was at every point where danger called; he gave the Gauls neither time nor means of uniting; he fought them successively, and beat them in detail. A few weeks sufficed to finish that remarkable campaign.

The study of the Commentaries of this great captain is one of the most instructive that can be recommended to young officers. Almost every page exemplifies the application of strategical principles, which, as has been remarked, are the same at all times and in all places.

Retreats.—In retrograde, as in forward movements, the simple line is the best, as by using it the mass of the force is always in hand to oppose the enemy. What are called *divergent* or *eccentric* retreats, which are effected at the same time by several routes, with the view of misleading the victor, and making him uncertain as to the course to pursue, are extremely dangerous. A force which is divided up to follow these different directions is weak at every point; the isolated corps, affording no mutual support, are exposed to the risk of being enveloped, thrust upon obstacles, dispersed, destroyed, as was the case with the Prussians after the battle of Jena. The victor, not allowing his attention to be diverted, should follow closely one of these corps and overwhelm it; he need

3*

give himself no present concern about the others, well knowing that he will have no difficulty in attending to them separately at a convenient time.

An army should never be divided up except just after a brilliant victory; it is then in presence of a disorganized enemy, who has lost his communications and is in a state of demoralization. It may rush into the midst of the scattered corps; it need only show itself and the enemy flies. In such a case no attempt is too rash; any thing is good except what is too slow and methodical. But this is an exceptional case, produced by the circumstances of the defeat of the enemy.

Remain united, therefore, in a retreat, still more than in an offensive movement, even if your motions are thereby retarded, for the first consideration is safety. March in as good order as circumstances permit, or at least keep together; still present an imposing front to the enemy, and, if he pursues too rashly, make him pay the penalty for so doing; have the boldness to turn upon him if he gives a fair opportunity or marches negligently. More honor is sometimes derived from a well-conducted retreat than in a battle gained, where chance often enters so largely.

Marshal Massena's retreat, in Portugal, in 1809, is an excellent model for imitation. He knew how to take advantage of all the accidents of the ground to retard the pursuit of the English. He never gave up a position until it was just about being turned, and then he fell back to another at some distance further on. His columns retired slowly, afforded mutual support, and kept out of each other's way, deploying and fighting whenever the enemy pressed them too closely, or where the ground was favorable for defence. Again they fell back, moving towards a common point in the rear, keeping always near enough for mutual assistance. Nothing, said an eye-witness, could exceed the skill there displayed by Massena.

It is very advantageous to retreat in a direction parallel to the frontier, if possible, because then an enemy gains nothing by his pursuit. This subject was touched upon at the same time with accidental lines of operations. If the *parallel retreat* is executed in an enemy's country, the army lives at his expense; the burdens of the war weigh upon him; he is almost as badly off as if he were the beaten party. If the retreat is made behind the frontier, the victorious army is drawn along after the other; it is forced to pass over much ground without advancing an inch towards the interior; only the borders of the country are given up to the pursuing enemy, which is also in a position where the flank is exposed to forces coming from the interior. But the retiring army should take care not to expose itself by undertaking such a movement in a country where there are no obstacles, for the enemy would then be able to cut it off from its base. The parallel retreat should therefore not be undertaken unless some respectable obstacle, as a river or mountain range, favors it.

If the parallel retreat is covered by a river, all the bridges should be broken, for security against attack on the flank. For the same reason, if it is made under cover of a chain of mountains, the lateral passes should be occupied until the army has gone by, and such disposition should be adopted as will permit an attack in force upon those hostile bodies which, notwithstanding the precaution taken, may succeed in breaking through or turning the passages, and may attempt to stop the progress of the retreating army. A weak corps presenting itself in this way should occasion no alarm, for the danger is really on its side.

Evidently a parallel retreat can only be effectively carried out on a frontier of considerable extent. If such a frontier is guarded with difficulty, on account of its length, it presents the advantage which has just been indicated, and this is not the only one, if the general knows how to adopt the active defensive system. For what cannot be undertaken in the defence of a narrow frontier without risk of being cut off from it, or being thrown back upon the obstacles which contract it while giving support to the wings, may be attempted with success upon an extensive frontier that presents long bases of operations. Those offensive returns may then be made which fortune often crowns with success, and those brilliant enterprises undertaken that are inspired by courage and daring.

Diversions.—Combined marches.—What has been thus far said to demonstrate the necessity of concentration of forces and of keeping united, proves also that diversions, combined movements, detachments, are operations that generally cannot be approved.

A *diversion* is effected when a corps is sent to a distance to operate independently of the army. A *combined march* is made when the movements of this corps and of the army have a mutual relation, and the same general object in view. In each case the separate corps forms a detachment.

Diversions are dangerous, because the army is weakened by just so many men as are employed that way; they withdraw the attention of the commander-in-chief from the main object; they increase the chances of accidents; they complicate events, multiply jurisdictions which may lead to clashing of orders, and they are almost always a source of disaster. The army, if victorious, can only gain a partial success, and can with difficulty profit by it; if, on the contrary, it is defeated, it is exposed to the danger of total ruin, since it can receive no support from the corps so unfortunately detached. However, there is nothing absolutely fixed in the difficult science of war; there is no rule without numerous exceptions. Cases arise, therefore, in which diversions are not only justifiable, but necessary. For example, a formidable position is to be carried by force, and it can only be done by attracting the enemy's attention in another direction; it then becomes necessary to detach a body of troops, more or less numerous, according to circumstances, who may make a circuit for the purpose of occupying commanding heights, or threatening the enemy's line of retreat. These cases frequently occur in mountainous countries; a detachment then becomes a necessity, the only alternative being the failure to accomplish the object in view; but it should be so regulated that its absence should be as short as possible, and, as soon as circumstances permit, there must be a return to the observance of the principle which requires unity of action and concentration of forces. Combined marches and diversions are much less dangerous in mountainous countries than elsewhere, because it is difficult for the enemy to interpose himself between the columns. The separate corps find in every valley contracted spaces where their wings are supported, and they are in no danger of being enveloped; and sometimes it would be even more dangerous than useful for large bodies to be held together. In such a case, each of the separate corps should be strong enough to defend the valley in which it is operating, and to keep open its communications to the rear. The rule is here violated only in form and not in reality, since there is no more subdivision than the ground requires; the different corps are not exposed to be forced from their positions, and there is ample opportunity for reunion by the roads in rear. General Lecomte gave a fine example of such marches when he attacked the Saint Gothard, in 1799. Upon that occasion, in all the valleys occupied by the French, they were in sufficient force to defend them even against a sudden attack of very superior numbers. That is the whole secret.

A diversion is allowable when the forces in hand are greatly superior to the enemy's, and there is difficulty experienced in subsisting them together. Skill is then shown in making such a division of the troops as will permit a corps to be moved against the flanks or the communications of the enemy while a force equal to his is presented in front; diversions may then be made in the provinces which are poorly guarded, or lukewarm in their allegiance, or where insurrection may be excited; troops may be unexpectedly sent to the capital or the richest cities, to levy contributions, &c. The detached corps should then act vigorously, make forced marches in order to multiply itself in the eyes of the enemy, cause him very great uneasiness or inflict a real blow. Indecision or vacillation on the part of the commander of the hostile army may induce a diversion. Except in these cases, it is much better to resist the temptation of diversions; it is always safest.

When diversions are condemned, it is to be understood that those are not referred to which may change the whole face of the war, or consist in an army's abandonment of its own country to the enemy in order to carry the war into his. They are altogether different from other diversions; the army remains united, and moves as a whole to the attainment of a single well-determined object. Such resolutions bear the true impress of genius, and, far from being blamable, are worthy of praise, no matter how they may result; for a man of spirit can attempt nothing more honorable for the salvation of his country after having tried in vain all ordinary means. Agathocles, king of Syracuse, was besieged by the Carthaginians; after exhausting all his resources in the defence of the place, being upon the point of surrendering, he determines boldly to pass over into Africa. He leaves in Syracuse only the garrison strictly necessary for its defence, takes with him his best troops, burns his fleet on the coast of Africa, so as to make victory a necessity, and advances towards Carthage. He overwhelms the opposing army, succeeds in contracting certain alliances, and brings the capital to the brink of ruin. Syracuse was saved. Certainly this was a diversion which produced a very great result; it would be a misuse of words to condemn an operation having the same name, though differing so essentially in character from those already referred to.

Diversions of this kind, although on a small scale, may be successful when they are well conducted and localities favor. It is always advantageous to do what the enemy is not prepared to expect, because in this way his combinations are thwarted, and he is obliged to stand on the defensive instead of taking the offensive. Turenne, in his last campaign, gives a fine example of such a course. He was not intimidated by the attack of his adversary, the celebrated Montecuculi, who had crossed the Rhine; but, taking no notice of his initiative, he crossed the river at another point, and obliged his opponent to leave the French territory in order to follow him and defend his own.

We ordinarily understand *combined marches* to be those arranged with the intention of arriving from two or more directions upon a position occupied by the enemy; or taking in front and rear an army which is to be attacked; or placing it, to use the ordinary expression, between two fires. But there is nothing more influenced by chance than these eccentric movements; independently of the temporary weakening they occasion, the smallest accident is sufficient to lead to their failure, and upset plans apparently perfect: a body of troops is led astray by the guide, bad roads retard the march, a storm arises, a swollen stream stops the column, the enemy is encountered where he was not expected; finally, a thousand accidents happen which cause the operation to be unsuccessful. On the other hand, the army may in the mean time have been attacked or forced to retreat; it may not be at the place of rendezvous; then the isolated corps is very much compromised; it is in danger of being enveloped and obliged to lay down its arms. The greater the extent of these eccentric movements, the more they are exposed to chance, and consequently the more care is to be taken to avoid them. Thus, on a field of battle, it is a fault to send a corps to the rear of an enemy to attack him there, or to cut his communications, because the army may be defeated while this corps is making its détour; but the fault is still greater when the detached corps is to make a march of several days, in order to reach a point of rendezvous already occupied by the enemy, for success then depends on circumstances that cannot be controlled.

Detachments made to effect a diversion or a combined march, or for any other motive, are condemned by all writers on the art of war. Many examples may be cited to prove their danger. The Great Frederic, usually so wise and skilful, had to repent having made, near Dresden, a large detachment of 18,000 men, with the intention of cutting the communications of the Austrian army with Bohemia. The detachment was surrounded, and fell into the enemy's hands after having fought bravely against triple its numbers, constantly hoping that the army would come up to its delivery. This affair took place at Maxen, in the month of October, 1759; it shows at once both the danger of detachments and that of wishing to cut the line of retreat of an army, which, although beaten, is still not entirely disorganized. The detachment, commanded by General Fink, succeeded, it is true, in taking a position in rear of the Austrian army and closing the way, but it was not strong enough to hold it. This general was certainly deserving blame for having permitted himself to be enveloped, for a force of that magnitude ought always to be able to penetrate through the enveloping circle at some point. Fink should have made the attempt, and he would have saved a part or the whole of his detachment.

Nearly upon the same ground, the corps of Vandamme, debouching from Pirna after the battle of Dresden, in 1813, had advanced to Töplitz in Bohemia, while the mass of the French army was still in the vicinity of Dresden. This detachment experienced at Culm the same fate as that of Frederic. However, Vandamme attempted to break through the enemy's line, and a part of his corps escaped in that way. Napoleon, in sending out this detachment, deprived himself of a part of his troops for the battle, or at least for the operations subsequent to the victory and having the object of completing it; he lost 10,000 or 12,000 excellent troops; and, what was worse, the *morale* of the army was sensibly affected by this check. If all the detachments referred to in history were as disastrous as these, the temptation to make them in presence of an enemy, unless he is entirely defeated, would be overcome; but others which are mentioned were entirely successful, and hence arises an attraction for them which it is difficult to resist. In fact, there is no more brilliant operation than cutting the line of retreat of the enemy. If, however, this can only be done by a division of force, it ought to be enough for a general of prudence to know the danger to which he exposes himself, to cause him to reject the idea of a diversion or of any separation of the different parts of his army.

Pursuit.—After a victory, the beaten corps, which are separated and disorganized by defeat, should be vigorously followed up; every effort should be put forth to prevent their rallying, and to take advantage of their temporary weakness and discouragement. There is no danger in dividing the army in order to give greater mobility to the columns intrusted with the pursuit, provided, however, they keep on interior lines and push the enemy on diverging lines. Care must be taken not to drive towards a common point corps already separated, because this would be playing directly into their hands and facilitating their rallying. A fault of this kind lost the battle of Waterloo to the French. Marshal Grouchy, who had been directed to pursue the Prussians after their defeat at Ligny, should have gained their right, in order to separate them from the bridges of the Dyle, which were their communications with the English army; instead of that, he pushed them towards Wavre, where they crossed the river, and came up at the decisive moment to the aid of their allies on the field of Waterloo. Grouchy, who should at all hazards have barred the way to them, or at least have reached the field of battle simultaneously with them, only followed them up, and did not make his appearance until the battle was over. Moreover, the terrible cannonade said in tones of thunder that the great question at issue was being decided at Waterloo, and that he should hasten there with his whole corps. In this memorable instance Grouchy exhibited a want of skill or of resolution; he did not know how to keep the Prussians at a distance, and, after permitting them to pass, he took no steps to neutralize the material and moral effect necessarily produced by their arrival upon the French army, which had been fighting since noon against superior forces.

When the enemy retires in tolerable order, his corps remaining united, and not presenting their flanks to partial attacks, the pursuit is made by the army *en masse.* It is necessary, however, to do something more than simply follow him along the same road, because, as soon as he finds a defile or other favorable position, he will check the pursuer, and compel the use of means to dislodge him ; precious time will thus be consumed, which may enable his re-enforcements to arrive. It is, therefore, judicious to manœuvre to gain the flank of the enemy, while a part of the army presses closely in rear. In this way he is unable to take any position that cannot be turned at once, and he will be forced to fall back immediately after establishing himself. The march of the pursuing army will thus be not perceptibly retarded, and the beaten party will experience the greatest difficulty in effecting a reorganization of his force. General Kutusof, in the disastrous campaign of 1812, took advantage of this method, called the *parallel pursuit.* Instead of following the same road as the French army, where he would have suffered much from want of provisions, and would have had continued combats with strong rearguards, he moved alongside the long column of the enemy, attacking whenever he could, throwing his forces into the gaps occasioned by the length of the road, by the cold, and by the sufferings of the French. This wise conduct on the part of the Russian general greatly increased the losses and demoralization of his enemy.

When the pursuing army has outstripped the enemy, there are two courses which may be taken; either to close the way, if it is sufficiently strong to brave his despair, or to take a position on the flank, leaving the road open. The latter is the better course, because it is always dangerous to impose upon any body of men, weak as it may be, the necessity of conquering or dying. Under such circumstances, men of spirit surpass themselves, and sell their lives or liberties very dearly; nothing is to be gained by placing them in such a situation. It is then better to attack such a body of troops by the flank than to bar their retreat; for they will seek rather to escape than to fight; a determination, however, which will cost them dearly. A general, expecting to occupy with impunity the path of an army or a strong detachment, must have forces considerably superior, to prevent its breaking through; otherwise, he should make a bridge of gold rather than expose himself to great loss, or even to entire defeat, as was the lot of the Bavarians at Hanau, when in 1813 they attempted to close the way to France to Napoleon's army, which was retiring after the disaster of Leipsic.

The passage of a river is always a delicate operation for an army which is retiring; the general engaged in the pursuit should know how to profit by this circumstance, to make an opportune attack. The advantages are evidently on his side; he attacks with his united forces an army divided by the river into two portions, which cannot assist each other; he surrounds the portion on the same side of the river with himself, and presses it upon an obstacle which should cause its destruction. Now or never he should act with the greatest promptness to take possession of the bridge, the only means of escape. Any hesitation under such circumstances would be a fault.

Guarding a conquered country.—It is not sufficient to overcome the enemy at every point, but arrangements should be made for holding the conquered country. For this purpose, detachments are made of sufficient strength to occupy the fortified places and other military posts, to collect the taxes imposed, and to keep the people in subjection. These detachments weaken the army in proportion as it penetrates further into the country; whilst the defenders, concentrating in the interior, are getting stronger. A point must therefore be reached where an equality will again exist, and there the struggle will recommence, the invaders having against them all the chances which may result from a defeat at a distance from the base of operations, and in the midst of a population ready to rise in arms at the reverse of their enemies, as was done in the Spanish war of Napoleon.

These inconveniences may be avoided by having the *active army* followed up by corps, whose only duty it is to guard the country. They will form an *army of reserve*, which will remain at the distance of several days' march, and will act according to special principles. Instead of remaining concentrated, it will extend itself as much as possible, in order to enlarge the base of operations, to procure abundant supplies, and to keep the country in subjection. The army of reserve, being weaker than the active army, and formed chiefly of new soldiers, will be entirely secondary to the latter, and will receive orders from the same chief; properly speaking, it is but a portion of the main army, which will be added to it when circumstances require.

The army of reserve will follow the active army in all its movements, keeping at a certain distance from it; will secure its rear, will guard the dépôts, and preserve free communications with the sources of supply; will besiege or blockade the fortified places, which the active army may thus neglect and pass by; it will guard and secure the line of retreat in case of disaster; will construct necessary works of fortification to improve the lines of defence, to cover the bridges and secure the dépôts; in a word, will do every thing the active army could not do, without scattering too much and delaying its march.

It is evident that such arrangements can only be made when there is great superiority on the part of the invaders in material and moral force. In the reverse case, far from thinking of forming an army or corps of reserve in rear of the active army, all the available forces should be assembled for battles; for the essential thing is to have decisive victories. When these are obtained, the country is held in subjection, either by re-enforcements passing through it to the main army, or by detachments of the strictly minimum strength consistent with safety, which, under the name of *movable columns*, pass through the country, and multiply themselves in the eyes of the inhabitants, by continual marches and counter-marches.

When an army, after long fatigues, should take some rest, it should be spread out for accommodation in the villages, and with a view to embrace such an extent of country as will furnish subsistence for men and horses. This is called going into *cantonments*. Evidently this step is only taken when the enemy is so distant that there is no danger of attack from him. However, the contingency of his advancing against the cantonments should always be provided for, by making arrangements for meeting him with a sufficient force. A point of assembly is designated for the troops in the event of an attack, and care is taken to have it so far to the rear, that the enemy may not be able to reach it before them. The more advanced corps retires on this point, those on the right and left move towards it by flank marches, and those in rear come up like re-enforcements by forward movements. Thus the concentration is effected by the shortest routes. To facilitate the movements, and when the cantonments are to be occupied some time, the roads are repaired, new ones even are made through forests and marshes, bridges are thrown over intervening streams, dikes are formed across ravines, &c.: no trouble or labor should be spared which contributes to the safety of the army.

The cantonments occupied during winter are called *winter quarters*. They differ from those just referred to only in being of greater extent. Natural obstacles should be taken advantage of to cover and secure them from incursions of the enemy. They are ordinarily established behind a river, which may serve as a line of defence.

We here conclude what we have to say on the subject of strategy with the remark, that if the principles of this elevated branch of the art of war are so simple that any one may not only understand, but discuss them, their application in practice is very difficult, and requires much sagacity and tact. The problem to be solved is an indeterminate one, admitting of many solutions; a thousand circumstances complicate it; the data are often, indeed always, more or less uncertain; and in many cases action must be taken upon no other basis than conjecture. There are many exceptions to the guiding rules. Events succeeding each other rapidly; unforeseen occurrences; difficulties in procuring information; motives unperceived by the mass of people, but imperiously influencing a commander, oblige him to modify plans perfectly arranged. If we add that time, a most essential element, almost always fails, it will be understood that none but superior men are capable of putting in practice this science, whose principles are contained in so small a compass. Let us, therefore, abstain from judging unfortunate generals with too much severity. Let us recollect that at the moment when they were obliged to act, they were not accurately informed, either of the force or the position of the enemy; that, having no certain information on this subject, they were necessarily reduced to a weighing of probabilities; that of many things which became perfectly well known after the event, they were entirely and necessarily ignorant; that, if they had known them, they would, doubtless, have seen as well as we what was best to be done; that, perhaps, again, the force of circumstances, which is irresistible in war, has obliged them to act otherwise. Let us, therefore, be moderate in judging others. Let us be indulgent, or rather just, towards those who are, in all probability, competent to be our teachers; let us not forget that circumstances beyond their control may have induced those acts which seem to us faults, and that good fortune is often at the bottom of the most glorious results.

CHAPTER II.

ORGANIZATION, ETC.

Art. I.—Composition of an Army.

To form an army, something more is requisite than a mere assemblage of men with arms in their hands; those men must be obedient to the orders of a commander, who directs their movements, and causes them all to act for the attainment of a common object. Without this, an army is little more than a tumultuous mob, where confusion and disorder reign supreme. If the numerous individuals composing it do not obey a single will, and do not act for a common object, no plan can be certainly undertaken or carried out. Discipline is, therefore, of prime importance to an army; an essential, indispensable condition for its existence. Thus those troops which have been brought to the highest state of discipline have at all periods, and among all races, had the undoubted superiority. By their admirable discipline, the Romans triumphed over all nations, and gained the mastery of the world.

Honor is the most certain promoter of discipline, especially in the militia. Punishments which are too severe irritate the soldier without reforming; they should be reserved for rare cases, where, to arrest mutinous disorder, it becomes necessary to use the most powerful repressive measures. It is by using persuasive means with soldiers inclined to be regardless of duty; by treating them humanely, even when punishing; by carefully avoiding humiliating them, by contemptuous and offensive language; by seeking, on the contrary, to inspire them with sentiments of patriotism and honor, that a chief can expect to form cohorts which will be magnanimous in victory and unshaken in the midst of reverses. These are the only means which can be relied upon to produce soldiers who may be trusted in critical moments. If they are insufficient to make heroes; if enthusiasm alone is the mainspring of prodigies of valor, like that at Thermopylæ, we may be at least sure of obtaining every thing that can be expected from veteran and well-organized troops.

A commander should, therefore, never use harsh and contemptuous language; he should particularly avoid violent reproofs of those subordinates who have failed in their duties; he is not excusable for giving way to passionate expressions; he will always congratulate himself for restraining his tongue; the soldiers, who are perfectly able to appreciate the offence, will give him credit for his moderation, and will pass their own judgment on the guilty parties, and will be themselves more disposed to obedience.

But there is another extreme into which officers of militia easily fall, and it must be carefully avoided. It is that excessive familiarity which lowers and throws discredit upon the individual who resorts to it, and renders him contemptible in the eyes of the very persons whose favor he seeks to gain. It is this culpable weakness which is the means of filling the ears of commanders with the talk of bad soldiers, who are always indulging in grievances and complaints. It is destructive of all discipline, as it seals the eyes of the officers to the faults and disorders which should be rigidly punished or repressed. A commander who is too familiar with his soldiers, loses all control over them ; he is exposed to vulgarities and want of respect from those whom he has accustomed to regard him as an ordinary companion, and to speak to him as such.

The man to whom the charge of a body of soldiers is intrusted, should avoid two extremes : the severity which alienates, and the excessive familiarity which breeds contempt. He will preserve a proper mean, if he knows how to be just while not severe, and to be kind to individuals while requiring rigid performance of duty. If no fault goes unpunished ; if general good conduct and gallant deeds receive praise and worthy rewards, the chief will be at once feared and loved by his subordinates ; his order will be punctually executed, and every effort will be made to merit his approbation. Finally, discipline will be perfect.

If it were possible to select the men of whom to make up an army, great stature should not be the only thing to be sought after. The Romans were not a tall race, and yet what nation has equalled them in military exploits? They accomplished such great results by joining to military qualities the civil virtues. Strength of body, great height, courage, are doubtless valuable things; but, to make the real soldier, these should be accompanied by other qualities no less precious, such as sobriety, patience under privations, honor, and, above all, a pure and ardent spirit of patriotism. Those virtues alone are sufficient to ennoble the military profession, which is too often disgraced by excesses of every kind, as hurtful to the nation it should protect as to the enemy who is the cause or the pretext of them. Therefore, the republics of ancient times exercised the greatest precautions in the selection of the soldiers to whom their destinies were committed. No one was admitted to the honor of bearing arms in the service of the country who was not directly interested in its defence. Our modern institutions do not permit us to hope for so excellent a composition of our armies, but something similar might be attempted in the service of militia; and, in the regular army, the custom of recruiting its ranks from the scum of cities might be abandoned, and replaced by a military conscription, against which, it is true, there has always been a great outcry; but still it presents the only means of having an army filled with men whose interests are the same as those of other classes of citizens.

After these preliminary remarks, which, it is hoped, may not be useless, I enter upon my subject.

A great army, under the orders of a general-in-chief, is composed of corps commanded by subordinate generals, each corps containing all arms of the service, that is to say, infantry, artillery, and cavalry, in the proportions determined by experience. It will be evident that the division of a large army into several corps is absolutely necessary, in order to render that huge machine capable of obedience to the various impulsions it is to receive, and of assuming the various forms circumstances require.

An *army corps* rarely contains more than 30,000 men, and often lower, even among nations who have the greatest numbers of troops. Such a command is a great burden, and few men are capable of managing it creditably.

An *army corps* is divided into a certain number of divisions of infantry and cavalry, each of which is usually, in France, commanded by a general of division, the army corps itself being under the orders of a marshal of the empire. In the United States army a division is the command of a major-general. The number of divisions in an army corps is variable, depending on the strength of the divisions and the corps. The other arms, engineers and artillery, enter usually by companies in the composition of an army corps. If, however, there are several companies of the same arms, as is usually the case for artillery, they may be joined under the orders of one commander.

A division is composed of several brigades, usually

from two to four; in the United States service each is commanded by a brigadier-general.

Brigades are divided into regiments, each containing two or more; regiments into two or more battalions or squadrons, according as they are composed of infantry or cavalry. We hence see that for these two arms the denominations are the same, except the last subdivision, which may therefore be regarded as the unit in the composition of armies, and is often used in expressing the strength of a corps, the number of battalions and squadrons it contains being stated. In the United States army the artillery has a nominal regimental organization.

One man could not attend to all the duties imposed by the command of an army. He has, as assistants, a certain number of officers who compose his staff. They transmit his orders verbally or in writing; they arrange marches and encampments; they make examinations of ground, and collect all possible information with reference to the position and movements of the enemy; they receive and communicate with flags of truce, and attend to procuring spies, &c.; they make condensed reports of the state of the command, from the reports of its several subdivisions; they make detailed inspections of troops, quarters, hospitals, &c., to be certain that every thing is in good order; they preserve the correspondence and records; they give descriptions of the parts of the campaign, make maps of fields of battle, supervise the different services, and see to the execution of regulations and orders; in a word, they are the intermediaries the general makes use of in setting in motion the huge machine called an army, and causing its various movements to be made in an orderly and suitable manner, without interference between the several parts. The great variety of details composing the duties of the staff makes it necessary to divide them into bureaus, and to assign special duties to each bureau. The superior officer placed at their head should possess the confidence of the commander; he must be made acquainted with his plans, in order to co-operate fully in carrying them out; it is his duty to give his opinions freely on the plans proposed, even to offer new ones if he thinks fit to do so; but it is equally incumbent upon him to throw aside his own ideas when they have not been adopted, in order that he may *imbibe*, so to speak, those of the general, as the least misunderstanding between these two men may lead to the gravest mishaps. Unity of ideas and action is the best requisite for success.

Each army corps, division, brigade, and regiment has its own staff, whose duties are similar to those of the staff of an army, although, of course, more contracted, and less important, proportionally to the body of troops in question. The artillery and engineers, when found in sufficient numbers, are assimilated to the divisions of infantry and cavalry. They have their special staffs for directing their troops, and every thing relating to their special and appropriate duties.

The administrative service of an army is usually divided into several distinct departments, as:

Pay department.		
Subsistence department.		
Medical	"	These two in United States service united.
Hospital	"	
Recruiting	"	
Clothing	"	These four in United States army combined in one, called quartermaster's department.
Barrack	"	
Fuel	"	
Transportation	"	
Military justice.		

These all have their employés, who swell the numbers of an army, although they do not enter the line of battle.

Thus, to recapitulate, a grand army is composed of several corps, which are sometimes designated as *the wings, the centre, the general advanced guard, the general rear-guard, reserve corps, &c.* Each army corps is formed of several divisions, each division of several brigades, each brigade of several regiments, and each regiment of several battalions or squadrons. The artillery attached to different corps, is usually formed into companies which serve batteries. They are kept together as much as possible. Sappers, miners, and pontoniers serve in companies.

An army is accompanied by a great number of carriages, for the transportation of munitions and warlike equipments, provisions, money, hospital stores, &c., tools, pontoons, &c., baggage, and a variety of other things. All these accompaniments were aptly styled by the Romans *impedimenta*, for nothing embarrasses the movements of an army so much as the long trains of wagons it is forced to carry with it, in order to supply its various wants. It is almost superfluous to say that these should be kept down to the smallest possible number, and the strictest care should be taken to prevent officers and employés from carrying with them carriages unauthorized by regulations.

The three arms, infantry, cavalry, and artillery, enter in different proportions in armies, according to the nature of the country where the war is to be carried on. In a mountainous country less cavalry is necessary, for there is but little ground suitable for its action; less artillery is admissible, especially of heavy calibres, because of the difficulty of transporting it. In countries full of plains, the artillery and cavalry should be considerably increased. The usual proportion, in the great armies of Europe, is for the cavalry to be one-fifth, and three pieces of artillery to every 1,000 men. Thus, an army of 100,000 men would contain 20,000 horses for cavalry, and 300 cannon.

The three arms are still usually distinguished as troops of the line and light troops: the first are specially designed to fight in order, with closed ranks; the others to scour the country, harass the enemy, pursue him after victory, make prisoners, &c. It would, however, be an error to suppose that these light troops are never drawn up in any kind of order, and never fight except as skirmishers. Such troops would be a great embarrassment. It seems still questionable whether two distinct classes of infantry, doing different duties, are necessary; an infantry of the line to fight in the ranks, and light infantry to act individually. Will it not be better to instruct all infantry soldiers in these double duties? A great simplicity would be the result; every body of troops would then be able to do its own scouting and similar duties; to spread out in broken ground, and to act in mass on level ground. The troops of the line would no longer be heard complaining of the absence of the light troops, nor the latter sometimes giving way on account of their separation from the infantry of the line. There would no longer be a pretext for the different corps accusing each other of want of success. The infantry would then, under all circumstances, be self-sustaining; being composed of similar elements, it would be able to form line of battle at pleasure, or to act in dispersed order.

When the duties are divided, one class of infantry cannot dispense with the services of the other; and if by accident they are temporarily separated, there is danger of loss of confidence; because the one is unable to disperse among woods and rocks, or the other, being in a level position, is unaccustomed to fighting in line. It has, moreover, been observed that troops of the line deteriorate in a long campaign, whilst the contrary is the case with light troops; all the infantry should therefore be required to perform in turn duties as light troops, and in the line of battle. Detached troops engage in what is called "*la petite guerre*"—an excellent school for officers, who, in such expeditions, must make use of all their faculties. Light infantry, seeming more suited to these duties, are preferred for them; so that the officers of the line, perpetually attached to their battalions, can learn nothing except the less elevating part of their profession. This is a very grave inconvenience, that would be avoided by giving all the infantry the same instruction. And certainly this would not be exacting too much of soldiers, who commonly pass a life of idleness in garrison, much more calculated to ruin their health than excite military virtues.

Certain military officers of great merit insist, notwithstanding these considerations, upon the propriety of having two kinds of infantry, in order that each may perform more perfectly the duty required of it, and have more confidence in itself. The principal strength of soldiers consisting in their own opinion of themselves, the soldier of the line will behave better in the closed ranks of a battalion, and the light infantry man in an open order which allows him full freedom of movement, where he can profit by his skill, and take advantage of the least obstacle offering a shelter. In fact, this question cannot be treated in a rigorous manner, as its solution depends upon national characteristics and varying circumstances.

As to the cavalry, it is best to have several kinds, in order to use profitably horses of all sizes, and to take advantage of their different degrees of strength and speed. Cavalry of the line, or heavy cavalry, are usually mounted on the strongest horses; light cavalry, composed of hussars, chasseurs, and lancers, make use of the smallest and most active horses; the dragoons are a kind of intermediate cavalry, manœuvring in line with the first, or skirmishing with the second.

The artillery of the line is served by men on foot, and the gait, usually, is the walk or the trot of the horses, in order that the men may follow. But the system is now adopted of mounting the gunners upon the limbers and caissons, which permits much more rapid movements. Light artillery is served by mounted gunners, the usual gait being the gallop; it manœuvres with the cavalry, being able to move with equal rapidity, and be supported by it. The artillery of the line regulates its movements by those of the infantry, and should be always supported by it. Besides, there is the heavy or siege artillery, which is necessary for the attack and defence of fortified places: it only follows the armies at a distance, or remains in the parks; it is only brought on the ground when needed.

Art. II.—Formation of Troops.

Since the introduction of the musket, to the exclusion of the rifle, in all the armies of the civilized world, infantry is formed in not more than three ranks, and often in two. With this formation, all three ranks can use their arms at once; with four ranks it would be impracticable, because the arm is too short. The men of the fourth rank would kill or wound those of the first. Even in the three-rank formation there is danger of this, especially with new troops. Opinion in Europe is divided as to the relative advantage of the formation in two or three ranks, and practice in European armies is also variable. The English and Swiss adopt the former. Among some of the people who use the three-rank formation, the third is specially intended for duty as skirmishers, so that there are really but two ranks firing in line.

The advantages and disadvantages of the two methods balance each other; examples of success attending the use of both may be cited. The English are certainly firm, notwithstanding their shallow order; the French deemed it expedient to form in two ranks at the battle of Leipsic, where they were obliged to occupy a very great extent of ground in order to fight against vastly superior numbers. Napoleon, in his Memoirs, seems to favor this formation, but proposes to increase greatly the number of file-closers, really making a third rank, whose duty it is not to use their arms, but to preserve order in the ranks.

To use the third rank in firing, the first must kneel, and this movement requires a command for its execution. Firing at the word of command is rarely practicable when the action is in progress, for the voices of the commanders are drowned in the noise of cannon, drums, cries of the wounded, &c.; the excitement of the combat makes it impossible for soldiers to give that cool and continued attention, without which a large number cannot discharge and reload their pieces together. The firing necessarily becomes "*by file*," which can only be executed by the first two ranks. The third rank thus becomes useless in firing, while one more man in each file is exposed to each bullet of the enemy. An attempt has been made, in the French service, to make the men of the third rank load their muskets and pass them to those in the second, who may thus fire twice in rapid succession, and the firing is in theory more actively kept up than in the two-rank formation. Experience has demonstrated the fallacy of this expectation, as confusion is produced in exchanging arms between the second and third ranks, loss of time takes place, and there is, moreover, a repugnance on the part of the soldier to part with his own and to use the musket of another. This repugnance is praiseworthy, for the soldier should cling to his musket, as did the Spartan to his shield. He should never be permitted to lose it; and if he cannot bring the whole of it from the field of battle, he should be obliged to exhibit pieces of it.

The losses of troops drawn up in two ranks, whether in line or column, are less when exposed to artillery; they occupy with equal numbers a greater extent of ground, which is very advantageous for outflanking the wings of the enemy, or for resting their own on obstacles which would necessarily be beyond the line if formed in three ranks. The first being able to occupy the same ground with fewer men than the second, stronger reserves may be held in hand, in order to act at the decisive point and gain the victory. Finally, troops formed in two ranks march with greater ease, and, for the same reason, are more quickly instructed and formed.

If there is a shock to be received, a charge of cavalry to be repelled, the formation of three ranks has undoubted advantage, because the third rank supports the other two and gives them confidence; its fire may be very effective against the horsemen in the later moments of the charge, when the men of the first two ranks, having come to the position of "*charge bayonets*," are bracing themselves to receive the shock. At this movement they naturally lower the head and lean forward a little, which diminishes their height somewhat, so that those who are behind are able to discharge their pieces without injury to those in front, particularly as it is necessary to aim high in order to strike men on horseback.

Cavalry, all other things being equal, have a much better chance of success against a battalion which is very shallow and extended greatly, than against one drawn up with a narrow front and with deep files. For this reason a small square is much safer than a large one: a little body of a few infantry often resists all the efforts of a numerous cavalry, because combined attacks are impossible against so small an object, which can be only assailed by a few horsemen at once, and these may be always driven off, if the infantry are cool and manage their fire skilfully.

When an action is prolonged and losses occur, gaps are formed in the line of battle, which, being constantly increased, may eventually compromise the safety of the troops, unless measures are taken to fill them; this effect is specially to be apprehended in two-rank squares, on account of the diminution of fire. The third rank may be very usefully employed in closing these gaps. It is really the first reserve of the combatants, and, viewing it in this light, a commander would be wise who, when the fight is not at close quarters, should place his third rank behind an undulation of ground, either standing, or sitting, or lying down, to prevent useless exposure. The third rank may be used in carrying off the wounded, whose sufferings and cries produce a demoralizing effect, which should be avoided if possible. Finally, if, during the contest, it is necessary to re-enforce the skirmishers, or to send suddenly a detachment upon the flank of the enemy, or to some point important to be occupied, it may be done, without disarranging the plan of battle, by employing the men of the third rank. Marshal Saxe condemned battalions too much extended, on the ground that they were not firm, and were good for nothing but firing, which, he said, was never decisive.

If the battalion is considered by itself, it is evident that where it exceeds certain limits it marches and manœuvres badly, either on account of the difficulty of moving its parts together, or of the impossibility of hearing the voice of its chief from one end to the other. A long line always wavers, and is more or less disjointed when marching in line of battle, which, of all the methods of gaining ground towards the enemy, is the most natural and simple. A battalion of moderate extent can march more easily and for a longer time in good order. The same number of officers is necessary for a weak battalion as a strong one, and they are more expensive for the state than soldiers. Two armies, one of 90,000 men, formed in three ranks, and the other of 60,000 men, in two ranks, would require the same number of commissioned and non-commissioned officers. Grade would be more valuable, and the subordinate officers would have greater importance, in the first case than in the second, because they would be proportionably less numerous. Any thing depreciates in value as it becomes common.

In a very mountainous country, like Switzerland,

for example, there are special reasons for preferring the formation in three ranks, which are drawn from the nature of the country itself. This country, even in the most open portions, is cut up by woods, hills, rivulets; there is seldom found sufficient space for deploying several battalions; there is no level ground not flanked by woods or other natural obstacles. Hence it is better for the battalions to have less extent and more solidity, in order to be less disordered in a broken country, and to close the open spaces better. These are in fact the only places where the enemy can penetrate—the only points where attacks of cavalry are to be met. The woods and broken ground may always be sufficiently defended by riflemen and skirmishers. Firing should not be the only dependence of the battalion; every thing else should not be sacrificed to facility of firing; care should also be taken to have the lines firm, and to provide means of giving the men that self-reliance without which they cannot be expected to offer a prolonged resistance to the attacks of troops that may be better disciplined and more numerous.

It has been proposed to retain all the advantages of the battalion formation in two ranks, for firing and executing certain movements with convenience, and to give also, in part at least, and under certain circumstances, the advantages of three ranks. This is to be effected by keeping two or more companies of a battalion specially for skirmishers, and when not thus engaged to place them in rear of the flank companies, in order to strengthen these points of the line, and protect them from attack by the rear or flank. If the battalion has a powerful attack to resist, these companies may form a third rank. In a simple square, they may also furnish a third rank to the four fronts, or form a reserve, to be moved to any point the enemy may threaten. In moving against the enemy in columns of companies or divisions, these companies may form small columns on the flanks of the main column, keeping abreast with the last division or a little in rear of it; they will thus serve to enlarge the breach which may have been made. These auxiliary columns may then turn to the right and left, taking in flank the adjacent parts of the enemy's line, if it still remains firm next the breach; finally, they may pursue in open order.

A formation in an even number of divisions is advantageous, as it permits a battalion to be divided into two equal half battalions, and each part to be manœuvred separately, if necessary. The proper extent of front to give a battalion in line is determined by the condition that the voice of its chief at one end may be easily heard at the other. This will fix the proper number of men for a battalion. A suitable front is found to be about one hundred and fifty yards, and, if in two ranks, the battalion will contain about six hundred men. It is an excellent plan to have each division formed of a single company, because, in actual engagements, the battalion is manœuvred in column of divisions much oftener than in company, and at such times it is very advantageous to have the men under the immediate orders of their own captains. When a division is formed of two companies, one captain is in the ranks, and a whole company under the orders of a man with whose voice the men are not familiar. This is a frequent cause of confusion at moments when it is most to be avoided.

Cavalry is drawn up in two ranks, which are quite as many as are necessary, and even then the men of the rear rank cannot use their arms effectively. Considering this point alone, it would seem that cavalry should be drawn up in but one rank; but the men of the second rank are by no means useless; for they support, press forward, and excite to greater effort those of the front rank. Gaps in a line of cavalry are even more dangerous than in a line of infantry, and the men of the rear rank are at hand to fill such openings. If we could expect to determine the effect of the shock of a body of cavalry, as in a problem of mechanics, by multiplying the mass by the velocity, we would conclude that the formation of cavalry ought to be deeper than it is actually, since the mass would be increased by making the number of ranks greater, the front remaining unchanged. But such a hope would not be realized; for no cavalry, however well it may be drilled, will form a compact mass at the end of a charge. The boldest men and the faster horses get in advance, and the wounded, timid, and badly mounted men fall behind, so that the shock, if a collision takes place, is rather that of successive and separate individuals than of the whole body. It appears, therefore, that two ranks are enough. The squadron cannot have so great a front as a battalion, because it would be less easily managed, and the voice of its commander would be drowned by the rattling of arms and the noise of horses. Experience has shown that it should have about half the length of the battalion, or about seventy-five yards. Allowing, as is usual, a yard to a man, the squadron will consist of one hundred and fifty men and horses ; and if the roll contains more men, it will be with a view of keeping the ranks filled to the proper number, and making allowance for absentees, who are generally numerous, since a horseman is a compound animal, consisting of horse and rider, and rendered unserviceable by accident to either of its parts ; moreover, horsemen are more used for duties out of their ranks than infantry.

For artillery the unit is the battery, consisting usually of four or six pieces. It is best to have the pieces in each battery of the same kind and calibre, as there is always liability to confusion and delay from mixing different sorts of ammunition. As a general rule, it is best to have the corresponding parts alike in any military body, and this applies to the *personnel* as well as to the *matériel.* Replacements of disabled parts is thus facilitated. In line of battle each piece of artillery of large calibre should be allowed about fifteen yards, and twelve for smaller. These distances are necessary for the free passage of limbers and caissons. Each piece should have its own caisson. The line of caissons is about fifty yards behind the pieces. This distance may be varied, and advantage should be taken of inequalities of the ground to cover the caissons.

CHAPTER III.

MARCHES AND MANŒUVRES.

THE business of an army is to march and to fight. By rapid and wisely directed marches a skilful general prepares the way for a successful campaign, reaps the fruits of victory, or escapes from pursuing and superior forces. The subject of marches forms, therefore, a very important part of the art of war.

Article I.—Rules to be observed in Marches.

Marches are of two classes—those made near an enemy, and those made at a distance from him. In the latter, convenience and comfort are greatly considered; each man may have ample room, so as not to be crowded by his neighbors; the road may be given up to the carriages of different kinds, so as to allow free passage; and the soldiers may march on the sides. Care should be taken not to move very large bodies together, in order that the troops arriving in a town may be more easily lodged and fed. When too many men are accumulated in a town, it is sometimes very difficult to provide promptly for their wants; those who are obliged to wait have just cause for complaint, and disturbances may take place. It is sometimes im possible to avoid these delays and discomforts, and soldiers should therefore learn to bear them patiently This is the touchstone which proves the good soldier. There is more merit in patiently enduring the inconveniences necessarily attending the painful trade of war than in braving death in battle.

When the body of troops to be moved to any point is very large, it should be divided up into detachments, and these started successively at intervals of a day or more, that they may be distributed along the road. If two roads lead to the same point, part of the troops will be sent by each, the time being so regulated that they will arrive successively at the point of junction. The columns should never cross each other, because a tiresome delay will always be caused to that one which must wait for the other to pass. Dangerous contests may even result from such meetings, if the staff officers who regulate the march are not very careful to arrange the manner in which the columns shall pass each other. Unless specially ordered to do so, no column will halt to allow another to pass it.

Soldiers cannot make an entire march without halting, especially when the distance is considerable. A halt at mid-day is necessary, and should be long enough to give men time to rest themselves and take some food. It is, moreover, right to halt a few minutes in each hour, and to allow no leaving of ranks at other times. A few trusty men should always be in rear, to gather up stragglers and prevent marauding. An early hour in the morning should be taken for starting, especially in summer; but men should not be deprived of necessary sleep; that between midnight and three or four o'clock is found to be best and most refreshing. Night marches are bad for the health of troops, as has been ascertained by direct experiment; they are very fatiguing to mind and body, and there is more straggling than in daylight.

For marches made in an enemy's country other arrangements are necessary. In such cases comfort and convenience must be sacrificed to security; necessary restrictions and labors must be borne in order to avoid all danger of surprise, and to be always ready to receive an attack. Negligence in this respect is inexcusable; for the enemy may at any time make his appearance when least expected.

The first rule is to march in column, and with as wide a front as possible, in order to have the column as short as possible, so that in case of attack it may be most speedily concentrated. If the object of the movement is a simple change of position, to effect an assemblage or concentration of troops before coming to blows; if, in a word, the march is not within the limits of the enemy's operations, some freedom may be allowed in the ranks, and the column may march *at full distance.* It should, however, be closed to half distance when the march is a *manœuvre,* that is to say, one of those movements which precede a battle, and are intended to concentrate the troops upon the important point, to secure the communications, to cover weak points, to deceive the enemy, to distract his attention, to divide his forces, to threaten his line of retreat, to make him uneasy about his dépôts, &c. These *manœuvre-marches* are so called because they are manœuvres made on a large scale, out of range of cannon, and have not for their object a simple gain of ground, as is the case with an ordinary march, but to reach a suitable position on the field where a battle may follow. They are executed in the immediate neighborhood of the enemy, and really under his observation. They should therefore be characterized by perfect order and great celerity.

When the battle is imminent, the troops should be prepared for rapid deployment, and formed in close column, by divisions, on the road, or at the side of it. This may be called the *order preparatory to battle.* There is but a single case where full distances should be kept in a manœuvre-march, and this is when the flank is exposed to the enemy. A column at full distance may form line of battle by a simple wheel of its subdivisions. These flank movements are always dangerous, and should be avoided as much as possible. Generally, the head of the column reaches the battle-field first, and the propriety of closing it up is manifest, with a view of facilitating and hastening the deployments.

A column on the march will always be preceded by

an advanced guard, which searches and explores the ground, opens or repairs the roads, if necessary, defeats the attempts of the enemy to surprise the column, or draw it into ambuscades, &c. It gives warning of the approach of the enemy, receives the first attack, and thus secures the main body time to prepare for battle. The advanced guard is preceded by small detachments. Other small bodies, called flankers, are sent off to the right and left, to pass around villages and clumps of trees where the enemy might be concealed, around hills bordering the road, behind hedges, through ravines, fields of grain, &c. The column has, in addition, its own flankers, especially when it is isolated, for something may escape the advanced party. It is far better to take too many precautions, than too few.

The Duke of Vendôme was more fortunate than prudent, when, at Luzara, in 1702, he came near pitching his camp in presence of Prince Eugene's whole army, which was in battle array, concealed behind a dike. His presence was not at all suspected, so that the advanced guard, having reached the ground, did not move further to examine the neighborhood, and the pitching of the tents, with its attending confusion, was about to begin, when an accident saved Vendôme's army. An aide-de-camp, whose duty it was to establish the camp-guard, thought it advisable to place a sentinel on the dike, which was very near. Reaching that point, he discovered the infantry of Eugene lying concealed behind the dike, waiting the signal for the attack; while the cavalry was in line of battle at a greater distance. He immediately gave the alarm, and the troops, who had not broken their ranks, were able to repulse the attack. Ten minutes later, Vendôme would have lost his army and his reputation.

In the wars of the French Revolution, an inexperienced republican general had under his command a long column of infantry, which was moving on a road bordered by hedges. There were no advanced guard, no flankers, and much negligence in the column, which was much elongated. Suddenly the Vendean chief, Charette, fell upon the flank of the column, cut it through, and dispersed it in a moment. Bravery was of no avail in such circumstances, and there was nothing to be done but to escape by flight. Such was the result of ignorance or imprudence upon the part of a commanding officer. A rear-guard is also necessary to close the march, prevent the disorderly conduct of stragglers, and guard against unexpected attack in the rear. The main body is thus surrounded by detachments which look to its security.

Wagons should not be mixed with the troops, because, in an accidental engagement, they would seriously interfere with the prompt assembling of the different corps; and, in all cases, they make the column unnecessarily long. The vehicles, therefore, should move in a compact and orderly manner, behind the troops, in two files, if the roads are sufficiently wide, in order to reduce, by half, the space they occupy, which is always considerable. The baggage should have an escort, to protect it from the partisans who may slip in upon the rear of an army.

Working-men should be distributed at the head of every column, to level obstacles, fill up ditches and ruts, repair bridges, or strengthen them when they are weak, &c. By pursuing this course the march of a column may sometimes be made less rapid, but never entirely stopped. These workmen should have with them several wagons loaded with tools, timbers, ropes, and other necessary articles.

Another thing, seemingly of little, but really of great importance, which demands attention, is the regulation of the pace at the head of the column, to avoid its becoming elongated too much. The poorest marchers, or the heaviest troops, may be placed in front, and, for the same reason, ox-teams may be put before the other vehicles. When the enemy is still distant, an interval may be left between the various corps of which the column is composed ; by this means the fluctuations of one are not transmitted to the others, and each moves with almost as much comfort as if alone. Otherwise, every man in rear is obliged to halt when, for any cause, those in front of him do so, and then he must quicken his pace to regain his proper distance. This irregularity of pace is found to be very fatiguing. It may be greatly avoided by adopting the plan indicated, but in manœuvre-marches, and whenever there is danger of an attack, distances must be carefully preserved.

Ordinary marches are from fifteen to twenty miles. The latter distance is a long march, but sometimes circumstances require a column to get over twenty-five miles, and even more, at the risk of leaving many men and animals behind.

Several days' rest should always follow forced marches, else the army will melt away in a little while, for the human body is not made of iron. At a decisive moment, and to obtain important results, a general may, and should, demand of his troops an extraordinary effort; but it should be of short duration. Two or three forced marches in succession are all that should be expected from troops inured to fatigue; but raw troops cannot stand even so much.

Infantry marches about two and a half miles an hour, not counting halts; so that, taking every thing into consideration, ten hours are necessary for passing over twenty miles, the march being in column. Cavalry makes about three miles an hour at a walk, and five miles at a moderate trot; as this gait may be held for several hours, it may, if necessary, pass over the space of an ordinary day's march in three or four hours; but forced marches are even more hurtful to cavalry than to infantry.

In marches near the enemy, the halt at mid-day is taken advantage of to cook. When men and horses have eaten something, they can with more ease get over the remaining part of the march, and have sufficient strength to fight, if an engagement occurs in the afternoon. Unless there are orders to the contrary, or special circumstances interfere, this halt should be long enough to give time for cooking.

The commissaries should be careful to have provisions ready at the end of each march, and, with this end in view, should have the column followed by animals and wagons loaded with provisions, procured by purchase, and by requisitions on the inhabitants, if necessary. These resources may sometimes fail, and it is a good plan to require each soldier to carry provisions sufficient to last him several days, prohibiting them, as far as possible, from eating unless when it is really necessary. In the Russian campaign, Marshal Davoust had arranged the knapsacks of his men so that they could carry in them four biscuits of a pound weight, and under each a little bag of flour weighing ten pounds, besides a cloth bag suspended from the shoulder, and containing two loaves, of three pounds weight. The entire load was nearly sixty pounds—but little less than that habitually carried by the Roman soldier. Whether the troops are in quarters or bivouacs, they should never break ranks until the detachments, whose duty it is, have examined the neighborhood, and the outposts are established around the quarters or bivouacs. Security is the fruit of vigilance. Every body of troops, however numerous it may be, should take these precautionary measures, not only at night, but at all halts.

Scouting Parties.—No march should be made in an enemy's country without a careful examination of the ground. In an unobstructed country this duty will be performed generally by mounted soldiers, but in a broken country by footmen. The scouting parties move in front and on the flanks. It is useless to have them composed of many persons, as their duty is not to fight, but to learn every thing they can. The duty is, moreover, a very fatiguing one, and should not recur too frequently.

The scouting parties precede the advanced guard. The parties on the flanks are called flankers. They move at two or three hundred paces from the advanced guard, of which and of each other they never lose sight. For this purpose the leading detachment sends out three small groups, one in front on the road, and two others on the right and left. Each group keeps together, or at least the members of it should always be in sight and hearing of each other. The flankers are sometimes obliged to move at a considerable distance from the road, in order to examine the country well. Each detachment would, in such a case, surround itself with small groups, one in front, and the others towards the enemy. When the country is much covered by forests, &c., the number of scouting parties must be increased.

Three men are enough for a group, the senior commanding the remaining two. They do not use their arms unless they fall into an ambuscade, or they are on the point of being taken, and then they must give notice of the enemy's presence in the only way which remains. We may here call to mind the noble devotion of the Chevalier d'Assas, who, having fallen into the hands of a hostile patrol, and being threatened with death if he uttered a sound, cried out at the top of his voice, "This way, Auvergne; here are the enemy!" He fell, covered with wounds, but he gave the alarm and saved his comrades.

The scouting parties endeavor to cover themselves under hedges, woods, thickets, or heights along the road, to see as much as possible without being seen. As soon as they discover a body of the enemy, they halt and conceal themselves, while one of the number gives the information to the commanding officer of the detachment to which they belong. They will keep him informed of every thing more they learn, and all without noise.

The scouting parties explore ravines and woods with great care, and will never pass a dike, a hedge, a wall, or a field of tall grain, without seeing whether there is any thing concealed. They should visit houses in which the enemy might be concealed, one of the scouts entering alone, while the others remain outside to give the alarm, if necessary. In like manner every spot should be examined where an enemy might be hid.

Before entering a village, the advanced guard halts, to give the scouts time to examine it, and procure information about parties of the enemy that may be near. The scouts pass through the streets, enter yards and gardens enclosed by walls, cause several houses to be opened to their inspection, and visit the churches and other public buildings. To take as little time as possible, the business of examination should be divided up, and ought to take but a few minutes. If the enemy is quite near, and there is any thing suspicious about the village, the scouts should take a longer time, and make a more thorough inspection. Moreover, however small the number of the advanced guard may be, it should be also preceded by an advanced guard, so that only the advanced detachment shall be delayed; and this body will retake its place by a rapid movement, after passing the village. In this case the advanced guard of the advanced guard would furnish the scouts, and the flankers would come from the advanced guard itself.

As the flankers are sometimes out of sight, connection is kept up with them by patrols or small detachments. The leading group on the road should consist of about five men, of whom two are sent in front of the remaining three.

Figure 6 gives the idea of these arrangements. A represents the advanced guard. B, the detachment that furnishes the three groups of scouts which precede it. C and C, the detachments which move parallel to the road, and furnish flankers, who are supposed also to be in three groups, the one in front of the detachment communicating with the scouts, and the other two groups on the outer flank. D and D are patrols, to keep up communication with the flankers, and affording them prompt aid if attacked.

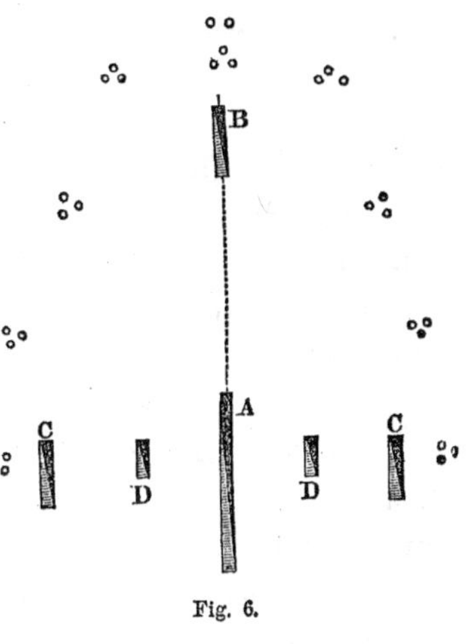

Fig. 6.

In a night march all these detachments should be drawn in much nearer to the advanced guard, and generally the flankers must be dispensed with. Silence must be kept, and every ear ready to catch the least sound. Xenophon says that at night ears must do the duty of eyes.

Advanced Guard.—If the only duties of an advanced guard were those of examination and skirmishing, none but light troops should be employed; but it has sometimes to seize and hold important points against a superior force until the main body can come up. It has to force defiles, take possession of villages, and is at every moment exposed to serious attacks. The advanced guard of a corps of considerable importance ought therefore to have, in addition to the light troops, some infantry of the line, heavy cavalry, and some artillery, in such proportions that the different arms may afford each other mutual assistance.

The battalions and squadrons of the army should all be detailed in turn for advanced guard duty, since this is the best school for soldiers of all grades to learn their trade, and all should receive its instructions. To form the advanced guard of detachments taken from the several corps is a bad plan, as it complicates the details of that service, and destroys the spirit of honorable rivalry which should subsist between the corps. Discipline suffers also, because the men do not serve under the chief to whom they are accustomed. For these reasons it should be laid down as a rule, that the battalions and squadrons should serve in succession on advanced guard duty. If the advanced guard does not contain a battalion or a squadron, whole companies should be detailed, and never fractions.

The commander of the advanced guard ought to have the best maps that can be obtained, and to be attended by several of the inhabitants of the country, to serve as guides, and give him such information as he may need. At each station he takes new guides, who are carried along under a strong escort, to prevent their escape. Knowledge of the language of the country is of great importance, and, if the commander himself does not possess it, he should have on his staff some officers who do, preferring those who have travelled through the country before.

Travellers, deserters, and prisoners, should be separately questioned as to the position and forces of the enemy, and with reference to what is known or presumed about his plans, the *morale* of his troops, the character of the commanders, &c. From each individual some information will be obtained, often indeed very incomplete and unsatisfactory, but by comparing all their statements a quite accurate idea may be formed of the composition of the hostile forces, their numbers, positions, prior movements, &c. All answers should be exactly written out and transmitted to the commanding general, whenever they furnish any interesting information as to the plans of the enemy.

As soon as the commander of the advanced guard arrives at the place where he is to halt, he questions the magistrates and other intelligent persons, in order to obtain all the information they possess about the roads, bridges, fords, forests, defiles, marshes, &c. If they know any thing of the position and designs of the enemy, they can generally be prevailed upon, by polite management rather than by threats, to communicate it. What they say with regard to the resources of the country is to be received with caution, for it is generally their interest to depreciate them. This, however, is a point about which reliable information should be required. While the commander is attend ing to these details, which demand much activity and tact, he sends some of his subordinates to examine the neighborhood, and specially in the direction of the road to be passed over the next day. The country should be explored to a considerable distance by patrols, which will move along all the roads, and inspect every spot where an enemy might be concealed. The troops should not lay down their arms, nor enter quarters, nor bivouac, until the patrols return and the guards are established, and there is no longer any thing like attack to be apprehended.

A very important duty of the commander of an advanced guard is to obtain provisions, in order to prevent his men from wandering off in search of them. As the advanced guard is generally not very strong in numbers, and may have to engage very superior forces, it is the more important for it to keep united. As provisions are needed to-day, so they will be to-morrow, and they should be loaded into some of the vehicles of the country. Such foresight is never hurtful. Montluc says: "Always have a good stock of provisions, and especially bread and wine, in order to afford the soldier some refreshment, for the human body is not made of iron."

When the advanced guard comes to a village on the road, it passes round it, if possible, rather than through it, especially if it is quite long, because defiles, where an attack may be made, are generally to be avoided. Sometimes, moreover, it may happen that the scouts have not made a sufficiently close examination of a village; the enemy may be concealed in the houses, and, under their cover, may do much mischief before they can be dislodged. Although this event is not very probable, yet it is possible, and the precaution is therefore justifiable and prudent.

At crossings of roads the advanced guard marks out with branches of trees or with straw the one it has taken, so that the main body, which is at some distance in rear, and generally out of sight, may not mistake the road. A few horsemen may be left at such a place to show the way, but that will depend upon the time required for the main body to come up. If a bridge is found broken or unsafe, or ditches dug across the road, or there are miry places or other obstacles which might stop the march of the column, the workmen immediately commence the necessary repair with all possible diligence, and rejoin the advanced guard at the first halt.

When the country is much covered with forests, the advanced guard sends out some scouts who keep up communication with the leading detachment and the flankers. This is to provide for the case of the leading scouts having left some portion of the ground without sufficient exploration, and the enemy, not having been discovered, might unexpectedly fall upon the advanced guard. The second set of scouts will see him approach and give warning of it. The same precaution may be used for the main body, especially when it is considerably distant from the advanced guard. We repeat, that in war it is impossible to take too many precautions to prevent surprise, as well on the march as in quarters.

In passing defiles that cannot be avoided, the leading detachment of the advanced guard should, after passing through, examine with minute care the neighboring ground, because it is in such positions that an enemy would secrete himself, with a view of attacking the column before it had time to deploy. The leading detachment should therefore halt after passing the defile, form line of battle, and send flankers out as far as possible, at the same time that the scouts examine all the ground in the front. Its march will only be resumed when the head of the advanced guard is at hand, and then its proper distance will be regained by increasing the pace. It is evident that this duty is very fatiguing, and consequently all the light troops of the advanced guard should take their turns, they alone being employed to form the leading detachment of the advanced guard.

There is no good reason for dispensing with any of these precautions in an open country, for however level it may appear, there are always undulations under whose cover troops may lie hid in considerable numbers, and an ambuscade in such a country would probably be the more effectual because the less expected. The only advantage presented by a level and open country is, that a few horsemen may gallop over it rapidly, giving it a sufficient examination, and thus the column be not long delayed.

If the advanced guard meets the enemy, it forms line of battle and tries to hold its position; if this cannot be done, it falls back upon the main body, taking advantage of all favorable points for delaying the enemy, and giving time to the main body to make such preparations as may suit the occasion. The commander should lose no time in sending an express messenger to the main body. If it is night, the advanced guard should charge with impetuosity, whatever may be its force, in order to throw the enemy into confusion or to intimidate him; at all events, to check him. In such a case the darkness is favorable to the attacking party, because the enemy, being unable to see what force he is engaging, must be careful. The main body can thus have time to come to the support of the advanced guard. But, as has been already observed, night marches are to be avoided, since they are very fatiguing, give rise to disorderly conduct which the officers cannot repress, and during the darkness most unfortunate mistakes often occur. One example will be given of the many that have occurred of friendly troops firing into each other at night. A French army in two columns was marching at night towards Landau, when a partisan chief with fifty men slipped between the two columns, which were separated by a ravine. He fired from his central position upon both columns at the same time.

Each, supposing itself suddenly attacked, returned the fire, and so they continued butchering each other until daylight revealed the fatal error.

The advanced guard should choose some favorable spot for the principal halt of the march—one concealed as much as practicable from observation, as, for instance, the reverse slope of a hill. Sentinels should be placed at the top to have an extended view, and upon the roads by which an enemy might approach. If there are roads near, they should be examined by scouting parties and a few men left there, who may conceal themselves in the undergrowth, and be enabled to see the exterior ground without being seen. If an attack is imminent, one-half of the troops will remain in line of battle, ready to fight until the remaining half take their meal. The halves will then change places.

When a wood of considerable extent is to be crossed, the advanced guard should be halted before entering it, and the leading detachment strengthened, in order that the number of scouts may be increased.

The drums of the advanced guard should not be beat, as they give warning of its approach; and generally the drum should be little used on the march, except in passing through a town, in coming into camp, or entering the field of battle.

What should be the strength of the advanced guard, and at what distance should it march from the main body? These questions cannot be answered with precision. The strength of the advanced guard and its composition vary with circumstances, and depend upon the means at the disposal of the commander-in-chief. The advanced guard may be temporarily re-enforced when it may have to receive an attack, or carry and hold a position; but in general its strength will not exceed one-fifth of the main body, and oftener will be less. The advanced guard will contain, in the majority of cases, between one-tenth and one-fifth of the main body. To make it stronger would be to fatigue the men by a severe duty recurring too frequently; on the other hand, to make it too weak would be to expose it to capture or destruction. A consciousness of weakness might prevent risking a vigorous effort when imperatively required, and such a failure might be highly disastrous to the army.

The distance of the advanced guard from the main body may evidently be greater for a large than a small body of troops, and it depends also very much on the length of the column. As it is the duty of the advanced guard to give notice of the approach of the enemy and to delay his march, it should be at such a distance that the main body may have time enough to prepare for an attack. For example, if the column is three miles long, more than an hour is needed for the rear to come up to a line with the head, and the advanced guard should, therefore, be at least the same distance in front; for if it is true that it delays the enemy's arrival, it is equally true that some time is consumed in ascertaining the real state of affairs and sending a messenger to the rear. It would seem then a good rule, that the advanced guard should precede the main column by a distance at least equal to the length of the latter. Frequently the distance is much increased.

The *rear-guard* is arranged nearly in the same way as the advanced guard, but is naturally made weaker, as there is less danger of attack in rear than in front. It is the escort for the baggage when the march is in advance. The rear-guard proper has its own rear detachment and flankers. These detachments send out groups of two or three men, who should frequently turn back to see whether the enemy is following the column. They will arrest deserters and marauders and drive along stragglers. A rear-guard should have a detachment of cavalry to move about rapidly in every direction when necessary, and to keep up communication with the main body.

In a retreat, the rear-guard becomes the most important body, and should be composed of the best troops, or those which have suffered least. There can be but little rotation in such duties. Necessity generally requires the same body to act some time as the rear-guard, often for several successive weeks. No other service can give more fame to a body of troops than in such a case as this, where it exposes itself to danger, privation, and toil, less for itself than for the remainder of the army. More than one commander of a rear-guard has made his name celebrated.

In a retreat, the baggage wagons are not placed between the main body and the rear-guard, but are moved as far towards the head of the column as possible, to secure them against attack, and not to embarrass the defensive arrangements.

Strength of a column on the march.—It has been stated, in a preceding chapter, that an army of considerable size is always divided up into several corps, which, on a march, form separate columns, and move along parallel roads. The army has thus not only more mobility, but also threatens the enemy simultaneously in several directions, keeps him in a state of uncertainty as to the real point to be attacked, outflanks him if he makes a stand at any particular one, and can be subsisted with much greater ease. The number of columns will depend on the magnitude of the army and the character of the country. Generally the number will be increased near the battle-field, to facilitate the necessary deployments. There is, however, a limit, below which it is best to keep the strength of a marching column.

Suppose, for example, a large corps marching along a single road. It will be composed of troops of all arms, in the proper proportion. There will be an advanced guard from seven to ten miles, or a half march, in front of the main body. Couriers will take about an hour to pass over this distance, and the commander of the main body will hence have about two hours in which to give his orders and to deploy his columns, upon the supposition that the enemy is greatly superior to the advanced guard, and drives it back upon the main body. Hence, the total length of the column should not be more than six or seven miles, in order that the rear battalions may have time to arrive at the head of the column to participate in the engagement. Experience and calculation demonstrate that an army of 30,000 men, including artillery and cavalry, and with infantry in three ranks, is about as large as can be properly marched along a single road. This fixes the magnitude of the *army corps*, as it is termed. If the infantry is in two ranks, the corps should not contain more than 20,000 men.

Art. II.—Forward Movements, and the Combats they lead to.

Order of march of a Division.—In order to apply the rules laid down, and to have definite ideas as to the details of a forward movement, we will take an example of a division of four brigades, subdivided as follows :—

16	battalions of infantry of 700 men each	11,200	men.
10	companies of sharpshooters	1,000	"
2	" " cavalry	150	"
24	pieces of artillery, in 4 batteries	700	"
1	company of sappers	100	"
1	" " pontoniers	100	"
	Total	13,250	"

This total does not comprise the division and brigade staffs, or persons employed in the medical, quartermaster, and commissary departments. These troops in line of battle will occupy about two and a half miles if deployed in a single line, and the infantry is in two ranks. The column, if the infantry march by sections, and each piece of artillery be followed by its caisson, will be also about two and a half miles in length, notwithstanding the absence of the advanced guard and flankers. The advanced guard will be two and a half or three miles in advance of the column.

The rear-guard, which escorts the baggage, will be about one mile and a half to the rear. If the division carries four days' provisions in wagons, as is usual when concentrating before the enemy, 400 wagons will be required, and the train will be more than a mile long.

The infantry is divided into four brigades, each of four battalions; and as the marksmen are to be spread along the front in case of an engagement, two companies will be attached to each brigade and two to the artillery. The infantry brigades will march in turn at the head of the column, and that one which is in front will furnish the battalions and sharpshooters of the advanced guard. In like manner, each battery marches in turn with the advanced guard, unless there are special reasons to the contrary.

The advanced guard will consist of the company

of sappers, one company of cavalry, two of sharpshooters, one battery, and two battalions of infantry. The proportion will thus be one-eighth for the infantry, one-fifth for the sharpshooters, one-fourth for the artillery, and one-half for the cavalry. It is too large for the artillery and cavalry, but the entire advanced guard will be about in the proper proportion to the whole command. Some cavalry soldiers are necessary with an advanced guard, even in a very broken country, to serve as couriers between it and the main body. Not less than one company should be with the advanced guard in this case. One platoon will pass to the front, and furnish scouts for the main and parallel roads, while the other will remain at the head of the advanced guard, forming the escort of its commander. The sappers will follow the cavalry, having with them wagons of tools, and other things necessary for the repair of roads and bridges. The battalions will throw out flank detachments, who will also keep up a communication with the cavalry scouting parties.

The remainder of the troops will march along the road, as well closed up as possible, and in such order that the column may readily form line of battle to the front. Generally the artillery should be in the centre, so as to be better protected by the infantry, and the sharpshooters will be on the flanks. The order of march of the advanced guard will therefore be the following:—

1st. A platoon of cavalry in front of the advanced guard proper, and furnishing the scouts;

2d. Another platoon of cavalry at the head of the advanced guard;

3d. The company of sappers with their wagons;

4th. A company of sharpshooters;

5th. A battalion of infantry, sending out one company as flankers on the right;

6th. A battery of artillery;

7th. A battalion of infantry, sending out one company as flankers on the left;

8th. A company of sharpshooters.

If this column deploys to receive the enemy, the two platoons of cavalry and the company of sappers will fall to the rear, to be employed as may be necessary, and the flankers will gather in upon the wings to support the sharpshooters.

The main body will march (two or three miles in rear) in the following order:—

1st. Two remaining battalions of the brigade which has furnished the advanced guard. A company of the leading battalion will march three or four hundred paces in advance, like a second advanced guard, which will throw to the several groups front and flanks;

2d. The entire second brigade, with sharpshooters leading;

3d. Three batteries of artillery together, each piece followed immediately by its caisson, and the whole preceded by their two companies of sharpshooters;

4th. The third brigade, with sharpshooters leading;

5th. The fourth brigade, preceded by its sharpshooters, having detached one battalion to the rear as rear-guard;

6th. All the baggage, reserve ammunition, forges, ambulances, bridge train;

7th. The battalion which forms at once the baggage escort and the rear-guard, the company of pontoniers and the remaining company of cavalry closing the column.

It would seem at first glance that the pontoniers ought to be in front, to prepare for passing rivers whose bridges may be broken; but their train would clog the advanced guard, whose movements should be prompt and unembarrassed. At the first reverse the train would be lost, and the army deprived of a most important accessory to successful operations. The pontoniers will only be employed in passing rivers of some importance, and in such cases their train is easily moved to the front while the army is halting, as it must do. The sappers of the advanced guard can make all necessary arrangements for crossing brooks and small streams.

The division will march as indicated, in advancing towards the enemy. There will be defiles to pass, bridges to force, rivers to cross, and, at last, preparations made for battle, when the enemy takes up a position to hold it. These cases will be treated in a general way, without particular reference to the division, whose order of march has been described.

Passage of Defiles.—A grave disaster may result from rashly entering a long defile when the enemy is near, even if he is retreating after a defeat. It is much wiser to attempt to turn a defile held by the enemy, than to trust to the chances of forcing it by direct attack at the expense of much blood. However, there are cases when the necessity is urgent, and, cost what it may, the passage must be forced. Under such circumstances, the first thing to do is to take possession of the lateral heights, a necessary preliminary to success. The heights must be assailed with infantry columns, whose strength will be proportional to the resistance and difficulties to be met and overcome. The enemy will be driven from the heights, and his position in the defile turned. The army will not attempt to penetrate the defile until the columns on the flanks have taken possession of the heights. If the defile is very narrow, there will be a single column in it; and in such a case there should be considerable intervals between the several parts of the column, in order that if the leading portion is driven back, the confusion will not be transmitted to those behind. When the valley is broad, and its sides gently sloping, other columns may move parallel to the main one, at levels between the bottom and top. These side columns should keep somewhat in advance of the main column on the lowest ground. This arrangement is based upon the supposition that the stream of water usually found in such locations may be readily crossed, so that the communication between the several columns may be easily kept up. It is always unsafe to separate the parts of an army by impassable obstacles, in presence of an enemy, who may attack and crush one while the other is little else than an idle spectator. If the stream in the defile is not easily crossed, bridges must be thrown over it at several points, or the whole force must remain on one side.

If the defile is short and not occupied in force, the advanced guard should take possession of it, the light infantry clearing the heights, and the heavy infantry passing through the defile with the artillery and cavalry. The artillery takes the lead, that it may not be masked if it becomes necessary to open fire. The cavalry falls to the rear, as it can be of little use in such a situation. The sharpshooters, if there are any, form small columns on the flanks of the artillery, or they may be replaced by companies of regular infantry.

When the defile is quite long, and strongly held, the advanced guard halts at the entrance and waits for the main body. When all the troops have come up, arrangements will be made as previously indicated.

The rules to be followed are the same for a large or small body of troops. The heights must first be carried, and the enemy driven from them, before a movement is made into the defile. This principle cannot be violated with impunity, as more than one example has demonstrated.

In high mountains, defiles are so easily defended against direct attack, that it is preferable to turn them, which is generally possible, as paths more or less circuitous almost always exist by which a force may be thrown upon the flank or rear of the assailed. The sense of security on account of the strength of a position sometimes puts vigilance to sleep, and the place is taken.

Sometimes the enemy occupies but one side of the valley. If the heights cannot be carried in such a case, the attempt may be made to slip along the other side some skirmishers, and even artillery, so as to take the enemy in flank, while the remainder of the troops attack in front. It was thus the French forced the defile of Calliano.

The defence is not always made in the defile itself. Sometimes an attempt is made to envelope and overwhelm the column as it debouches. The commander-in-chief, having been informed of such a state of affairs, hastens to join the advanced guard, and, ascending some eminence whence he can see the enemy's position, he makes his arrangements and sends his orders to the troops in rear.

Generally, the only course is to scatter the troops just in front of the defile by a fire of artillery, and immediately to throw forward and deploy a portion of the troops on the ground thus cleared. These troops, under cover of the artillery crossing its fire in front of them, endeavor to gain ground to the front to make room for those behind them. The skirmishers, and especially the sharpshooters, push forward on the right and left along the heights, with a view of annoying the batteries of the enemy, whose converging fire on the debouching columns may be very destructive. In the mean time, new battalions, closed in mass in the defile, but at some distance from each other, debouch and deploy rapidly to the right and left. Several squadrons gallop up and deploy on the wings. When troops enough are on the ground, they attack with the bayonet and drive the enemy off.

In the deployments subsequent to passing defiles, and in other cases which might be mentioned, much time is often gained in manœuvring by *inversion.* Many instances have occurred of troops being repulsed because they did not know how to manœuvre in this way.

Whatever be the circumstances of the passage of a defile, and whatever plan be adopted, the baggage should never be permitted to enter until the outer extremity is clear of the enemy, and there is ample space for the troops to manœuvre. It is not difficult to conceive of the horrible confusion of a column driven back from the front, harassed in flank, and finding the defile filled with wagons, whose drivers have cut their harness and escaped on the horses. The baggage should therefore never be permitted to enter the defile until it is perfectly safe.

The passage of a defile of some importance requires, on the part of the advanced guard, redoubled vigilance and precaution, even when the enemy is supposed to be absent, for he may return by a cross-road and attack unexpectedly. The defile should, therefore, be examined by a detachment of cavalry some distance in front of the advanced guard, and the woods on the right and left carefully explored by flankers. The advanced guard should pass through by detachments, so that if any accident happen to the leading one, the remainder may escape the same fate, and may render it assistance. All the detachments should unite at the débouché, and remain there until scouting parties have thoroughly examined the neighboring ground, so that they may feel secure against all danger of attack.

Passage of Bridges.—Bridges are short defiles, but their passage requires some precautions to be taken by any body of troops. They present serious difficulties when they are defended, and the passage must be forced. When there is no enemy visible, the same precautions are necessary as for any other short defile. The leading detachment of the advanced guard stops at the river-bank, taking advantage of all cover afforded by the locality, until the scouts have made an examination of the ground on the other side. The horsemen, who perform this duty, cross the bridge at a rapid trot, and then divide into three groups; one of which moves carefully along the road, making the circuit of houses, gardens, and clumps of trees, while the other two groups go to the right and left, exploring as they go. When, after a few minutes, no enemy is discovered, the leading detachment crosses the bridge, and rapidly rejoins the groups which belong to it. In the mean time, the advanced guard has halted at some distance from the bridge; the flankers, having reached the river-bank, move down towards the bridge, examining as they go. They cross the bridge and move to the right and left, resuming their duties as flankers. The advanced guard then crosses the bridge, but always at a rapid pace, and in close order. It is a good plan to leave an interval between the artillery and the troops before and after, and to pass that over at a trot.

As soon as the advanced guard is safely over, the commander makes a report of the facts of the case to the commander-in-chief. When the main body comes up, it may pass without delay, only giving a little time for its flankers and scouts to pass before, and take their appropriate places on the other side. If the communication established by the bridge is an important one, the advanced guard should not move forward until the main body has come up, because the enemy might permit the passage by the advanced guard, and dispute that of the main body.

In such a case, it should prepare to resist any attack the enemy might make before the arrival of the main body, taking advantage of the features of the ground to strengthen its position.

When the enemy holds the bridge, it is very difficult to dislodge them, especially if the bridge is long, and the opposite bank the highest, and the enemy provided with artillery. The famous passage of the bridge of Lodi, by Napoleon, in the presence of 10,000 Austrians, is well known.

Artillery plays an important part in the attack of a bridge, because it is necessary to drive the enemy from the other side before sending troops upon the bridge. If the direct attack fails, the enemy's attention must be engaged while the main body passes at another point.

If the bridge is short, and is held by a small force, the advanced guard may be able to carry it. If not it will halt, the staff officers will examine the ground, look for fords, &c., while the main body is coming up.

The artillery will be put in position to cross their fire on the ground near the other end of the bridge, and marksmen may do excellent service also in killing the men and horses, if the river is not too wide. While this firing is in progress, the infantry, who are to force the passage, are making the necessary preparations, covering themselves as well as possible behind inequalities of the ground, houses, woods, or thickets near the bank. Each battalion will form a closed column, with as wide a front as the bridge will permit. The first battalion advances past the line of marksmen, rushes upon the bridge, and crosses it at a run; drives away the few men remaining at the other end, and deploys near the bridge. If this is successful, a second battalion crosses in like manner, deploys alongside the first, and the two gain ground to the front. A third battalion is thrown over, and finally the whole army is gotten across. But it may happen that the first battalion is repulsed. If so, the officers rally it, and again rush upon the enemy. If there is any hesitation in the movement, it is a fit time for a personal effort. The chief seizes a flag, takes his place in front; the men are inspired with fresh courage; the bravest rally around their leader; the rest follow, and the passage is forced. Evidently there must be some powerful motive to justify so dangerous and bloody an operation, else it would be preferable to take more time, and secure an easier passage at some other point.

As soon as the infantry has taken up a position on the opposite side, the artillery continues its fire as long as it can without endangering the infantry which has crossed. It then ceases fire, forms column, passes the bridge as rapidly as possible, and takes position on the other side, at some suitable point where it may continue to fire upon the retiring enemy. The cavalry will have crossed as soon as the infantry was in sufficient force to give it support, and is threatening the flanks of the enemy. Or else, taking advantage of some ford, it may have turned the position while the infantry were attacking in front. The remaining troops, who, during the engagement, have kept out of fire, and made such dispositions as would be advantageous in case of a repulse, now form column and cross the bridge.

Passage of Woods.—When a column is to pass through a large forest, the attempt should not be made until it has been thoroughly explored, both in front and on the sides, to such a distance as to remove all apprehension of unexpected attack by a large force of the enemy. In this case, the infantry must do the scouting, as the cavalry can make but little progress in a dense forest. The flankers should be supported by several patrols, which will keep up their connection with the column. If the wood is of small extent, the advanced guard does not pass through until the scouts have reached the other side; but if of considerable extent, the delay would be too great, and, in this case, a halt of half or quarter of an hour is made, while detachments follow each other to the front, and serve as successive re-enforcements to the leading detachment, if it is obliged to fall back.

If the wood is occupied by the enemy, skirmishers are sent forward, who gradually creep up on the flanks, under such cover as they can find, to the outskirts. The artillery follows at a suitable distance, takes a central position, fires obliquely, and forces the defenders to take refuge in the denser parts. Thus, the skirmishers are enabled to get cover behind the outer fringe of trees. When this is accomplished, and the advanced guard is in possession of the skirts of the wood, the enemy can not hold his position long, unless abatis have been made. This species of obstacle cannot be of very great extent, and may be the more easily turned, as the wood itself conceals the movement. The enemy, once set in motion, should be vigorously pressed by the skirmishers, supported, if practicable, by small columns. If the entire wood can be turned, the cavalry, which can be of no use in the interior, may make a circuit to threaten the communications of the enemy, and hasten his retreat. When the forest has been well swept by the fire, several squadrons may be sent along the road at a gallop, to dash upon the retiring enemy as he debouches into the open ground, and thus change his retreat into a rout.

Passage of Rivers.—The most serious obstacle that can be encountered on marches is an unfordable river, with its bridges broken down, and the enemy in possession of the opposite bank. To pass such an obstacle always occasions loss of time, which will be proportional to the difficulties attending the attempt. If the column has a bridge train, it will answer for the passage of streams of moderate size; but when the river is very broad, it may become necessary to collect the boats of the country and other materials for building the bridge.

A passage is made either by a stratagem or by force, or oftener still by both together. The attempt should not be made until the staff officers have made an examination of the river, to discover the most favorable point. This point will usually be one where the bank is higher than on the other side, and at the same time envelopes it, on account of a bend in the river. The neighborhood of an affluent is convenient, as it offers facilities for collecting in safety and floating down the bridge materials. A thickly wooded bank and a wooded island enable the preparations for passage to be made out of sight of the enemy.

Efforts are made to keep the plans secret, and to deceive the enemy. Preparations are sometimes made at a place where no passage is intended, and, when every thing is ready, a sudden move is made to the real point. At dawn, batteries, advantageously posted upon commanding points, on the right and left, cross their fire on the opposite shore, and drive off or under cover all who may be there. Marksmen may also be very useful in assisting the artillery, if the stream is not too wide. At the same time, the boats descend the affluent where they have been concealed. They are filled with soldiers, and rapidly moved across the river, where the men disembark as fast as possible, and secure the best cover they can, to enable them to hold their ground. The artillery gives all the aid it can render while the boats are carrying over other loads of infantry. There is a double advantage in these successive trips of the boats, the first soldiers thrown over being incited to strenuous efforts by having their retreat temporarily cut off, and still having the hope of speedy re-enforcement. The position of the attacking force gets constantly better until they feel strong enough to charge upon the enemy, and drive him from the river.

The artillery is silenced as soon as its fire might be dangerous to the troops on the other side, and will shift its position to another, where it may still be effective.

When things are in this condition the bridge may be commenced, but troops should continue to cross in the boats. Field-works are at once begun, to cover the bridge and secure its possession in case a retreat becomes necessary subsequently. This is a prudential measure which should never be omitted.

The bridge should be built above the affluent, so as not to be exposed to dangerous shocks from boats which might get loose and float down the current. When the bridge is finished, the artillery, the cavalry, and the remainder of the infantry pass over.

Arrival of the column in presence of the enemy.—As soon as the general receives notice of the presence of the enemy, he hastens to join the advanced guard, in order to examine his forces and position, as well as the ground upon which he must himself manœuvre and fight. He takes with him the chief of artillery, and at least one of the higher officers of each of the corps in the column, so that, after the reconnoissance, when his plans are arranged, he may give to these officers, in person, instructions for their respective corps.

In the mean time, the troops composing the column halt in the road, after sending forward several battalions to re-enforce the advanced guard, if necessary; and, while waiting orders, they close in mass, to occupy as little depth of ground as possible. The baggage, under guard of a few men, remains in rear, and the cavalry, which had formed its escort, files past along the road and takes position behind the infantry. The field artillery is in the centre of the column, and the reserve caissons draw near, to be at hand in case of need.

In this order the column will advance, either along the road, or moving off obliquely, if required. If the engagement is not to take place the same day, the army, upon reaching the position selected, will deploy, light its fires, and wait for the next day. When the general sends word to prepare at once for battle, the column will halt and break up into several smaller columns, which will move to the right and left at distances suitable for deployment, opening roads, if necessary, with the aid of the axe and shovel.

To have definite ideas on this matter, let us recur to the example of the division on the march, mentioned above in this chapter. The two remaining battalions of the leading brigade of the column will be rapidly moved forward to support the advanced guard. There will remain three brigades on the road in close column, the sharpshooters and eighteen pieces of artillery being between the second and third brigades. The order assumed preparatory for battle will be that

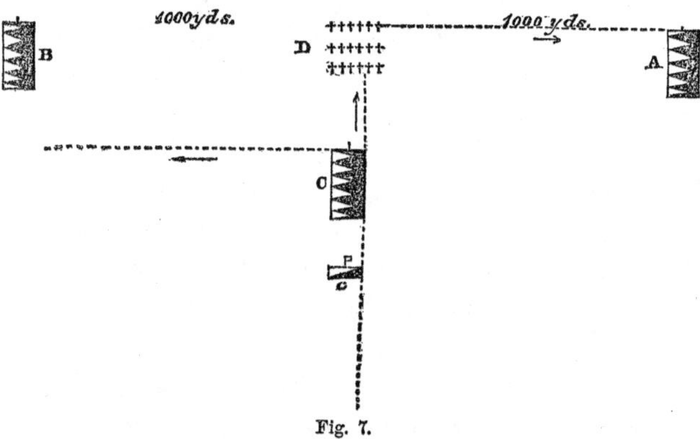

Fig. 7.

shown in Fig. 7, where A, B, and C are the brigades, D the artillery, and C the cavalry, and the division will move forward in this order. When it has arrived on the field, the necessary deployments can be made in twenty minutes. The subject of battles will be treated in another chapter.

Art. III.—Flank Marches.

As a rule, a flank should not be exposed to the enemy; but there are cases where a flank march cannot be avoided, or where it is the best thing to be done, and the risk should be run for a short time, in order to gain some great advantage. Take the case, for example, of a corps which could effect a junction with the main body, by executing a flank march along the position of the enemy, or by making a wide détour at a distance from him. If it was *essential* that the junction should be effected in the least possible time, the risk should be taken, and the flank march made. In war there is no rule without exceptions. These flank marches are exceptional cases, which sometimes occur, and it is proper to mention the precautionary measures to be adopted to avoid disaster on such occasions.

As the greatest danger to which the column is exposed is that of being attacked in flank during the march, a strong detachment should be thrown out on the side next the enemy, to move along in a direction parallel with the column, and sufficiently near to keep up a constant communication with it, and avoid being itself compromised. A mile would be about a proper distance between the column and the flank detachment, but this must depend upon the ground and other circumstances.

The advanced guard usually becomes the flank detachment on a flank march, but it will be well to increase its strength, as the chances of attack are greater, and it should be able to hold its ground against a serious effort of the enemy. The flanking detachment should have its advanced guard, rear-guard, and flankers, each of these sending out groups and single men in every direction, to avoid surprise. The main body will need but a small advanced guard to march a few hundred yards in its front, the ordinary rear-guard following also, but nearer than usual.

The baggage becomes, in a case like this, more of an encumbrance than usual. It may be sent to join the army by a circuitous route, at a distance from the enemy, or may move on that flank of its own column which is safest against attack.

Such are the principal arrangements necessary in a flank march. See Fig. 8, where M indicates the en-

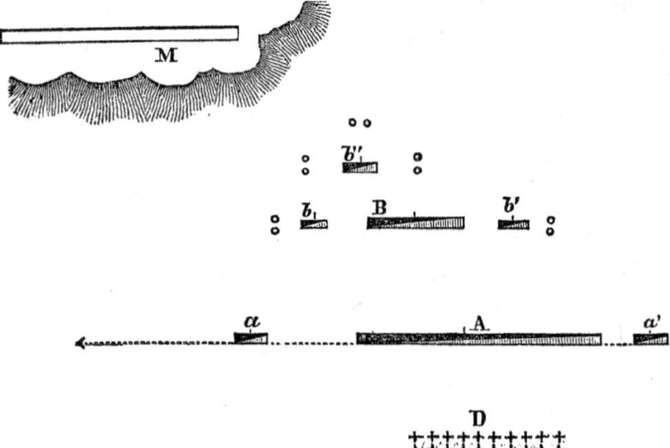

Fig. 8.

emy in position; A the principal column, with its advanced guard *a*, and rear-guard *a'*; B the flank detachment, with its advanced guard *b*, rear-guard *b'*, and detachment *b''*, on its flank. The scouts are only placed on the side next the enemy. The baggage, D, moves on a road parallel to that occupied by the principal column.

The principal column and the flanking detachment should each march with full distance between their subdivisions, so that, by simply wheeling to the right or left, as the case may be, line of battle is at once formed. Perfect order should be kept. If the flank detachment discovers a lateral defile, through which the enemy might debouch, such as a ridge, or a road through a marsh or village, that would not need a strong guard, a detachment should be left until the column has passed, and it will then join or replace the rear-guard. This would be a useless precaution in an open country, as small detachments would not then answer for such purposes, and, moreover, there is not the same liability to surprise as in a broken country.

It is obviously wise to conceal a flank march, if possible, under cover of a fog or the darkness of night, or any feature of the ground which would screen the movement from the enemy. Under no circumstances should the use of the flank detachment be omitted. If it encounters hostile patrols, a bold appearance must be put on, so as to lead to the belief that the whole army is at hand. In the mean time, the column moves steadily forward, and, when it has passed sufficiently far, the detachment leaves the position it had held, and retires rapidly, taking, if necessary, a direction different from that of the main column, and subsequently rejoining it by a détour. The enemy will not follow it very far, as he exposes his own flank to the troops which have already passed.

In flank marches, it is particularly necessary to have accurate information as to the character of the road, and the obstacles to be surmounted, for the least delay might prove fatal, if unexpected. If a defile is to be passed, troops should be sent forward to hold it.

If the flank march is to be executed by a large body of troops, several columns should be formed, if the ground is of a nature to allow them to move freely. The column at the greatest distance from the enemy should be also farthest to the front, the others being thrown back in echelon so as to afford mutual support. If the enemy attacks the first column, he is liable to a flank attack from the others; if he attacks the last column, those in front are in close supporting distance for it.

In Figure 9, M M represents the enemy in position; A is the flanking detachment; B, C, and D the three columns into which the mass to be moved is divided. The columns are moving left in front, so that by simply wheeling to the right, line of battle is formed in echelon towards the enemy. The baggage E may follow the same road as D, or a still more distant one,

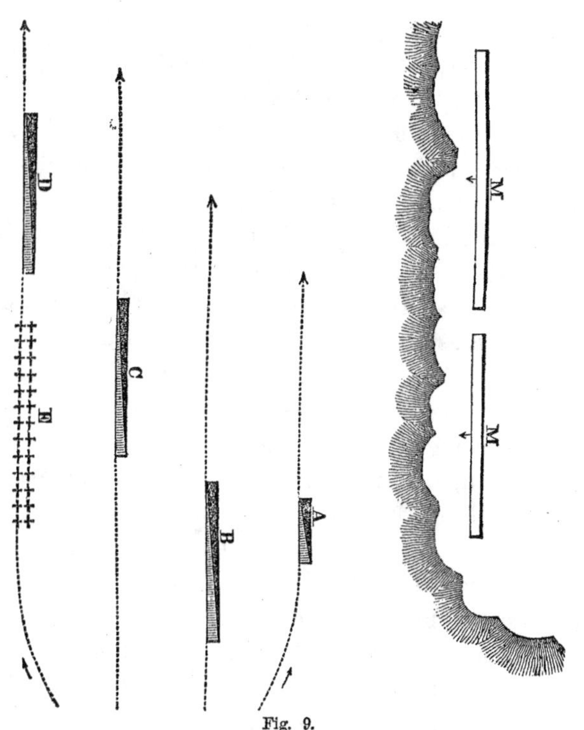

Fig. 9.

if there is such. The columns may be a thousand paces from each other. The first column might be too seriously compromised if the others were more distant.

When several columns are to march in this way by the flank, it may be necessary to make roads to enable them to preserve the proper distances. It was formerly a very common thing to make roads for the numerous columns of an army as they approached the field of battle; but in later years, when armies are moved with so much more rapidity and ease, it is seldom done. However, Napoleon had several roads cut out of the rocky banks of the Saale, in the night preceding the famous battle of Jéna. It might be done again when necessary.*

It was the habit of Frederick the Great, when his enemy was incapable of rapid manœuvres, to fall upon him by a flank march. He formed his own army on such occasions in two columns of companies, so that by wheeling to the right or left he at once formed line of battle facing the enemy. For this purpose, he changed the direction of his march when near the enemy, under cover of some inequality of ground and protected by the advanced guard. These long columns ran the risk of being attacked in front, and often would have been if the enemy had been more prompt, but Frederick knew with whom he had to do. Sometimes, to remedy this grave objection, he moved forward in four columns, and upon changing direction formed two which could wheel at once into a double line, constituting his line of battle. See Figure 10. The two central columns A and A are composed entirely of infantry; one-half of each goes into the front line, and the other half into the second line. The columns B and C, of cavalry, are also each divided equally between the front and second

* Roads were cut at Cerro Gordo by the Americans under General Scott.

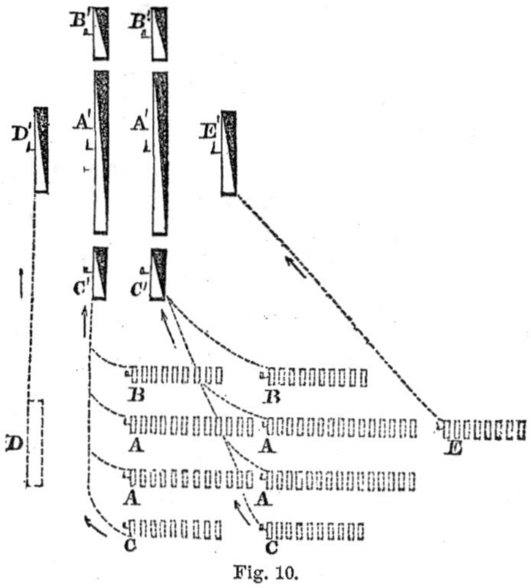

Fig. 10.

lines of the line of battle, of which they form the wings. An advanced guard D covers the heads of the columns and masks their movements. The rear-guard E of the columns, becomes the reserve of the line of battle.

Art. IV.—Marches in Retreat.

A retreat may be simply a retrograde movement of one army before another, without an engagement between them, which demands nothing more than the ordinary precautions for a march; or it may be after a defeat, and the general who conducts such an operation will need all the firmness and experience of the finished soldier. The essential thing in a retreat is to get the start of the pursuing enemy and shake him off. Forced marches become necessary, and they must often be made at night. Night marches are not objectionable in such circumstances, as there is little or no danger of the retreating army falling into an ambuscade. On the other hand, the pursuing army has that danger to apprehend, and must be proportionally circumspect. The necessary inconveniences and fatigue accompanying night marches must be expected and endured, for safety has become the paramount consideration. Marshal Turenne, after his defeat at Marienthal, held his ground until night, and then took advantage of the darkness to move off, completely distancing the Imperial troops, who dreaded an ambuscade and preferred to wait for daylight. Sometimes a movement to the rear may be concealed from the enemy by lighting the camp fires as if to spend the night, and leaving a few men to keep them burning, while the army passes off quietly. The detachments thus left behind may set off just before daylight and readily rejoin the main body.

When a start is gained in this way, it is important to keep it, even at the sacrifice of some of the wagons that move too slowly. In such times the safety of the troops is the important thing. Every thing, however, should be carried off that can be, and especially the artillery should be saved.

As soon as the commander has determined to retreat, he should order the officer having charge of the train of wagons to move off at once on the route indicated to him, and to push along as rapidly as possible. The road may thus be cleared to some distance to the rear.

It is a very difficult thing for an army to commence a retreat in presence of the enemy, especially after a severe battle. It should hold its own as well as it can, repulsing the attacks of the victorious army, and yielding ground only foot by foot, until some position is reached where a stand may be made during the remainder of the day. The general should take advantage of the first opportunity to collect his scattered troops, cause them to take some food and rest, while he arranges the order of march and organizes a rear-guard. Notwithstanding the darkness, each corps must take the place assigned it, after procuring the means for transporting the wounded. It will thus usually be midnight before the retreat really begins. If it is absolutely necessary to abandon the wounded, they should be collected in such houses as may be at hand, with physicians and other attendants, and left to the generosity of the victors.

The rear-guard should be formed of the best troops, or of those which have suffered least. It should be well provided with artillery, as this will keep the enemy at a distance, and force him to frequent deployments which delay his movement forward, and give the retiring army time to get a good start, and occupy some advantageous position for checking the pursuit. The rear-guard is expected to make a bold front wherever the ground is favorable; and as the main body continues its marches, the interval between them must often be considerable. The composition of the rear-guard should, therefore, be such that it may take care of itself on all kinds of ground. It is impossible to say exactly what should be its strength; but, as a general rule, a rear-guard in retreat should be stronger than an advanced guard in an advance. Less than one-fifth the whole force would be too little, and sometimes one-quarter would not be too much for the rear-guard, for, independently of its daily combats, its duties are extremely fatiguing, since the same troops, on account of the difficulty of changing them, are often obliged to constitute the rear-guard during the whole of a retreat.

On the march, the rear-guard is divided into three parts: its main body, which keeps together as much as possible; the rear detachment, which sends word of the approach of the enemy, and gives time to make arrangements to receive him; and an intermediate detachment, to keep up communication with the army, and occupy defiles which are to be passed. In addition, there are small detachments of flankers.

Perfect order should be preserved in the rear-guard. That detachment which is in the immediate presence of the enemy should march *by the rear rank*, so as to be with its proper front rank to the enemy, if obliged to halt and engage him.

In an open country, the rear-guard should march

in several columns. In a defile, on the contrary, the whole must be in a single column, except the flankers, who try to move along the heights. Squares are formed in open ground, if the cavalry of the enemy is annoying. These squares should afford mutual protection, by flanking each other with their fire. The rear detachment should be of light artillery, and the best cavalry in the command. If it becomes engaged with overwhelming forces, it must retire at a gallop, and take shelter with the columns of infantry, or in the intervals of the squares.

In a very broken country, infantry will form the rear detachment, as neither cavalry nor artillery could manœuvre freely. The cavalry and artillery would then be mostly with the leading portions of the rear-guard.

The commander of the rear-guard should be constantly on the *qui-vive*, to avoid being surprised. On the march, he will not only keep out his flankers, but will send out detachments, to occupy any points where the enemy might debouch upon the flank of his line of march. These detachments will remain in position until the whole rear-guard has passed, and will then join the rear detachment. At halts, the troops should be in order of battle, all approaches being carefully guarded. The men will not be all allowed to eat at once. The march will only be resumed when all the detachments have come in.

Passing a Bridge in Retreat.—If there is a bridge on

the road, the main body, after passing it, will leave a sufficient force to hold it until the arrival of the intermediate detachment, of which mention has been made. When the rear-guard is informed of the facts, it halts at some distance from the bridge and deploys, in order to check the pursuit of the enemy. When the resistance has been sufficiently prolonged, the rear-guard retires in echelon or in the checker order, and passes the bridge by the wings, taking position on the other bank, where a part of the artillery has already been advantageously placed. The rear detachment of the rear-guard makes a final effort to defend the entrance to the bridge, while the sappers make preparations for burning or otherwise destroying it. Several pieces of artillery will be placed in position to enfilade the bridge, and others will cross their fire in front of the entrance. The sharpshooters on the bank endeavor to retard the establishment of the enemy's batteries, and the battalions take advantage of all cover afforded by the ground, to protect themselves from the artillery fire.

The difficulties of the passage are much increased, if the bridge, from peculiarities of the ground, is opposite one of the extremities of the line of battle instead of the centre, as the rear-guard must then manœuvre by but one wing, and the other is necessarily fixed in front of the bridge until the remainder of the force has passed.

If the bridge is destroyed, the enemy is necessarily

checked for a greater or less time. If it cannot be destroyed, the rear-guard, after holding its position as long as possible without compromising its safety, will resume its march to the rear. The commander will first send to the rear those battalions that are farthest from the bridge; then a part of the artillery will take up some position where it may sweep the road and the neighboring ground, and the cavalry will be on the flanks, to charge the first troops engaging in the pursuit. Skirmishers will cover the retreat, taking advantage of ditches, hedges, and other like obstacles. They thus form a screen which protects and masks the columns.

It may be sometimes advisable to cross the bridge at a run, but care should be taken to avoid confusion in crossing, and there should be prompt return to the ranks, and order on the other side of the stream.

Passing other Defiles.—The hints given above will suggest the precautions to be taken in passing other defiles than bridges, such as a road through a marsh, a long, narrow mountain pass, a village with narrow streets, thickly built. In such cases, care must be taken to get well the start of the pursuing enemy, to look for, seize, and hold, until all have passed, any lateral roads by which he might attempt to attack in flank, and cut off a portion of the column. Narrow defiles may be obstructed by the use of various simple devices, and the enemy forced to great delay. It must not be forgotten that a defile is a favorable situation to form an ambuscade for part of the enemy's force. An attempt of this sort may often be carried out with ease, and produce the effect of entirely stopping the pursuit.

The commander of a rear-guard is often placed in the most trying positions. To acquit himself with credit, he should possess much firmness and activity, and be acquainted with all the resources of the tactician. He should possess the confidence of his men and the respect of his enemies. If there is no post more dangerous than that of commanding the rear-guard, there is certainly none more honorable. Marshal Ney acquired an imperishable fame by his conduct as chief of the rear-guard of the French army, in its sad and terrible retreat from Moscow. For several weeks successively he was fighting the enemy every day, and more than once engaged in hand to hand conflicts, like a simple grenadier. Soldiers will never forget the glorious deeds of that heroic spirit, who was the last to leave the hostile soil, after enduring, with a few brave comrades, all the dangers and privations poured by cruel fortune upon the heads of that famished and shattered army, which entered Russia in such imposing array but a short time before.

Art. V.—The simultaneous Movement of several Columns.

Thus far we have considered but a single column or army corps, marching along one road. It remains to say something with reference to the movement of a great army of several corps, upon the supposition that the enemy is near.

As each of these columns is liable to attack separately, and may have to take care of itself for a time, until supported, the same precautions for its safety should be taken as if it were isolated and alone. Thus every thing contained in the preceding articles is applicable to each of these columns. Each should have its advanced guard, rear-guard and flankers; and should break into as many smaller columns as circumstances may require, and the locality permit. Each should select its positions, encampments, or bivouacs, of course within the limits properly falling to it. But each should deport itself as a part of the great mass, and all manœuvre towards the common end.

Independently of the special advanced guards of the several columns, one of the corps should precede the others, and form a general advanced guard for the army, and another will form a general rear-guard or reserve. The arrangement of a great army of five corps, marching on three roads, is shown in fig. 11. The central road is occupied by three corps, A, B, and C, the first forming the general advanced guard, the second the centre of the army, and the third the general rear-guard, more properly in such a case called the reserve. The other two roads are followed by the corps D and E respectively, which serve as flankers to the principal column.

The lateral extent of ground, as D E, embraced by the columns, is called the front of the march. It is usually a distance of several leagues, so that all kinds of ground may be found within its limits. There is one rule never to be disregarded: to avoid leaving between any two columns obstacles such as rivers, lakes, extensive marshes, impassable ranges of hills, which would prevent the corps from communicating with each other, and rendering aid in case of need. If such an obstacle occur, either the whole army should make a détour to avoid it, or the column which would be separated from the others, if not moved in the original direction, should turn to the right or left and fall behind the nearest corps, where it would remain until enabled to regain its proper place. The corps on the flanks should occupy any defiles through which their communication is kept up with the main column.

The distance between two of the parallel columns should not be more than from five to six miles. The general advanced guard, being on the principal road, along which it may fall back, if necessary, should be

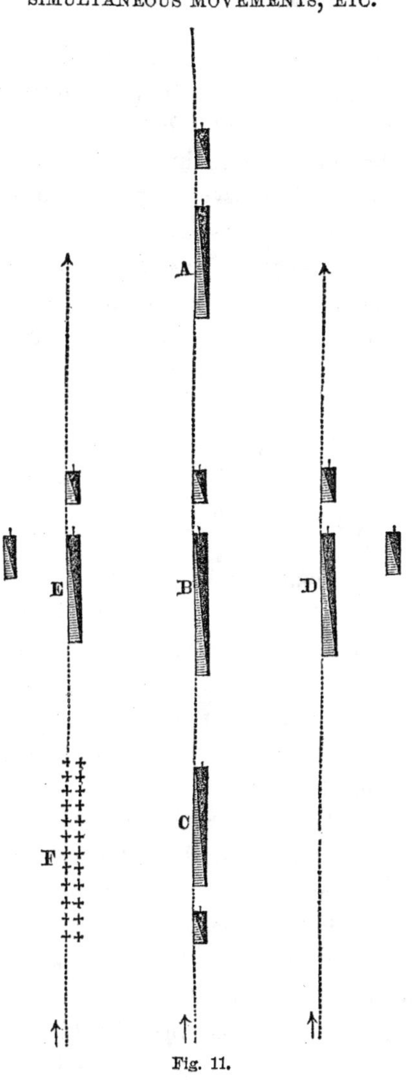

Fig. 11.

at a greater distance; in fact, it should be as far to the front as possible, consistently with safety. Twenty or twenty-five miles would often be not too great a distance, although a precise limit cannot be laid down for it. The general reserve may be from six to ten miles in rear of the main body.

The reserve artillery, the great train, and the most cumbrous baggage-wagons are collected, as well as possible, in the safest place, as at F in the figure. Each column, however, is accompanied by whatever is indispensable in the way of baggage, &c., which follows in its rear, under charge of a guard. The fewer the wagons, &c., attached to a column, the better for it; but there are always too many.

The commander-in-chief generally stays with the central corps, because he is then in the most convenient position for transmitting his orders to the other corps, and receiving reports from them. He may join the advanced guard when it is necessary to examine the ground before the army reaches it, or to see the disposition of the troops of the enemy.

In order that the march may be executed in the best manner, the chief of each corps ought to know its general object; to be told what corps are respectively before and behind him, and on his flanks; from what quarter he is to expect support in case of need; in what direction he must move if assailed by very superior forces. There should, moreover, be a constant interchange of couriers, orderly officers, or aides, between the general head-quarters and the several corps, in order that, on the one hand, the chief of the staff, who prepares the orders of march, may be constantly informed how they are carried out, and may communicate to the commanding general the real condition of affairs at any moment; and, on the other hand, that the commanders of corps may be kept acquainted with passing events, and instructed as to the necessary modification of details in carrying out the main plan. The orders sent to the different corps of the French army, by Berthier, during the marches which preceded Napoleon's great battles, are worthy of careful study.

If an enemy is met in front by an army marching, as in fig. 11, the advanced guard A falls back nearer to the main body, at least until it reaches a position that may be held, when it makes an effort to maintain it. The corps D and E oblique towards the central column, still gaining ground towards the front. The central corps B hastens up to the support of A. D and E unite and form a second line. The reserve C will come up and be employed as the commander-in-chief may think fit.

These are the most natural dispositions, but they may be modified in many ways; for example, D and E may join A, and form a first line, the other joining B, and forming a second line. Three corps may be put in the first line, and two in the second; or, two in the first and three in the second, &c. This will depend on the ulterior designs of the general, and the relative facilities offered by distances and the ground of bringing up one or the other corps to the support of the advanced guard; or he may be influenced by some other of the many circumstances which are constantly arising to interfere with what would be invariable rules in war. There is one great advantage in this, for the enemy cannot know, until the last moment, the disposition of the troops he is to contend against.

If the enemy shows himself in force in such a position as to threaten the flank of the army, all the corps change direction towards that side, and it will be seen that the general arrangement remains unchanged. There is still an advanced corps, a reserve, a central corps, and one on each flank. Suppose, for example (fig. 11), the army to have changed direction to the right: D forms the advanced guard, B remains in the centre, A and C are on the flanks, and E is the reserve. In retreat, or to attack an enemy who is in the rear, the arrangement is the same.

No general ever understood better this mechanical part of the art, or applied it more skilfully, than Napoleon. The history of this wonderful man offers to every student of military affairs the most instructive lessons. It is necessary to go back in history as far as Julius Cæsar to find a general at all to be compared with him.

The march of the Russian army before the battle of

Austerlitz was made in a manner entirely similar to that shown above. This army, under the orders of Kutosof, was moving along the road from Olmutz to Brunn (see fig. 12), in order to engage the French army, whose advanced guard was at Wichau. The French main body was between Brunn and Austerlitz. Kutosof divided his army of 95,000 men into five columns, leaving between them only space enough for deployment. The columns on the flanks were but subdivisions of the corps E and D, being under the same commander. Three corps were marching on the main road; A, commanded by Bagration, being the advanced guard; B, under the immediate eye of Kutosof, but having its own commander, a lieutenant-general, forming the centre; C, commanded by the Grand Duke Constantine, as the reserve. The two columns on the right, D, composed a corps under General Buxhöwden, and the two left columns, E, also a single corps, directed by Prince Lichtenstein. So far, all was well. The Russians were defeated, but it was because they committed the fault of making a long circular movement around Austerlitz with a portion of the army, in order

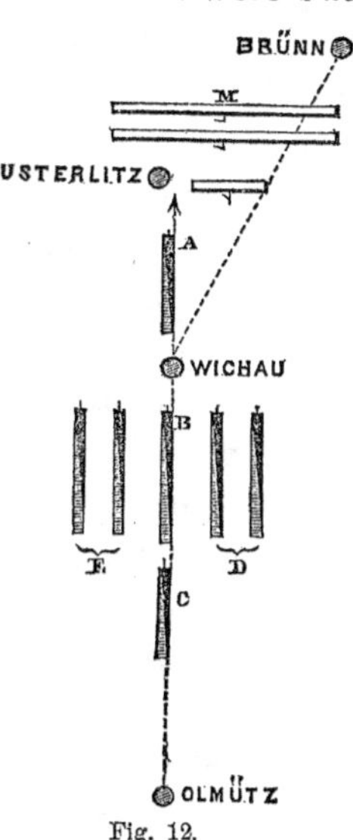

Fig. 12.

to turn the right of the French army, M. The centre was too much weakened: Napoleon fell in force upon that point, and the Russians were completely beaten.

After laying down the general rule, it must be stated that circumstances may, and often do, render modifications of it necessary. The essential thing is to conform to the principle, that the different corps, into which an army is necessarily divided, ought always to have such relative positions that their concentration may be easily effected, before the enemy can unite his forces against one of the corps. In an open country the rule must be strictly adhered to; but whenever any natural obstacle delays the march of the enemy, and renders a concentrated attack upon his part difficult, while, at the same time, protecting the movements of the other army, there may be greater space between the columns of the latter, and a more extended field may be embraced in the manœuvre, if some important result may thereby be gained, or the object of the campaign be more perfectly attained.

In 1805, when Napoleon had concentrated his army in the neighborhood of Stuttgard and Ludwigsburg, intending to cross the Danube and take in the rear the Austrian army, which was established near Ulm (see fig. 13), he divided his army into four columns, of which the first, on the right, commanded by Marshal Ney, was to move upon Günzburg, near Ulm, to attract the attention of the Austrians in that direction. The second column, composed of the corps of Soult, Murat, and Lannes, on the same road, and at a half day's march from each other, was to be directed upon Donauwerth, twenty-five miles lower down the Danube than Günzburg. The third column, containing the corps of Davoust and Marmont, passed the Danube at Newburg, ten or twelve miles below Donauwerth. Finally, the fourth column, under Bernadotte, was directed towards Ingolstadt, eight or ten miles lower still. There were bridges at all the towns named in the figure. In this manœuvre-march there was a distance of forty-five miles between the right and left columns, but the fact of a great river like the Danube being between the two armies, the certainty that the enemy had not collected his entire forces, made the probability of attack very small. If the right corps had been attacked, it would have fallen back upon the central column moving along the Danube, which would have been a protection to one of its flanks. The enemy, having his attention drawn to this retrograde movement, would have abandoned the right bank of the river,

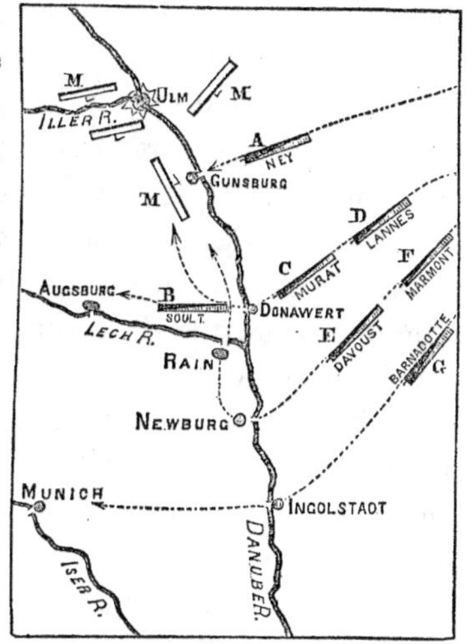

Fig. 13.

where was his line of operations, and the general manœuvre would have equally succeeded. There was, therefore, no real disadvantage in thus separating this column from the others. If, on the contrary, the enemy should make a flank movement towards Donauwerth, with all his disposable force, he would find there Napoleon himself and the principal French column, while he was all the time outflanked by the other two columns, the nearest of which might be brought up in a day. The distances were so calculated that the strongest column, the one by Donauwerth, was to arrive in advance of the others, and they were echeloned in such a manner, that the first having passed the Danube during the 6th and 7th of October, the second would arrive at Newburg on the 8th, and the last at Ingolstadt on the 10th. So that really only the Donauwerth column was attackable, and as it was composed of three corps near together, it could dispute the ground until the arrival of the others; or else, if forced to retire, it might move down the right bank of the Danube, protected in flank by the river, until reinforced the next day by the column from Newburg, and still later by the column from Ingolstadt, while, in the mean time, the Günzburg column was threatening the flank and rear of the enemy. The more study is given to this manœuvre-march, the wiser does it seem. There is no supposition to which it does not adjust itself, and still the front of the march was more than fifty miles. This is another proof that in war every thing depends on the skill with which the arrangements are made to suit the topography of the field of action.

After crossing the river, all the corps changed direction to the right, in order, by a movement of considerable extent, to envelope the position of Ulm, and take possession of the bridges of the Lech, by which the Austrians might attempt to escape. The whole army described a circle around Günzburg while Ney held that point, his corps becoming the pivot of that grand movement. Soult's corps, which at first formed the advanced guard, became the left flank; Murat and Lannes, who were in the centre, became the general advanced guard; Davoust and Marmont now formed the centre. Ney was on the right, and Bernadotte was the reserve. In this way the army, which had arrived upon the Danube, presenting the right flank to the enemy, advanced upon him by a front march, after passing the river. The results of this skilful plan are well known. Ulm capitulated; nearly a whole army was captured; only a small part of it escaped along the left bank of the Danube, while the French were passing to the right.

Fig. 13 exhibits the features of this combination. M M represents that part of the Austrian army which was assembled at Ulm and Günzburg; A is the corps of Ney flanking the other columns, and moving upon Günzburg, where he was to hold the bridges. B, C, and D, the corps of Soult, Murat, and Lannes, forming the principal column, are represented as arriving at Donauwerth; the first of these, having crossed the river the previous day, is marching towards Augsburg, to secure the bridge over the Lech at that point; the other two, after passing through Donauwerth and seizing Rain, turn to the right and form the new general advanced guard. E and F, commanded by Davoust and Victor, will reach Newburg the next day, and, having crossed the Danube at that point, will turn to the right and move towards the bridge of Rain, to form the centre of the army. Bernadotte's corps, G, making the fourth column, is directed upon Ingolstadt and Munich, being intended to be the general reserve for the army.

Enough has been said to explain the merely mechanical part of manœuvre-marches made by several columns, and to show the spirit of such combinations. If the country is open, and the enemy near, the columns keep together, and march so as to be ready to give prompt mutual support. If, on the contrary, the troops of the enemy are scattered or at a distance, or if large rivers intersect the country and give security against sudden attack, the columns may be separated by greater distances, and may even present themselves obliquely, provided their relative positions are preserved, so as to permit concentration in case of need. When shall the rule be strictly observed? When and how far may a deviation from it be permitted? These questions admit of no answers. Circumstances, which are always different, must decide in each particular case that arises. Here is the place for the general to show his ability. The military art would not be so difficult in practice, and those who have become so distinguished in it would not have acquired their renown, had it been a thing of invariable rules. To be really a general, a man must have great tact and discernment, in order to adopt the best plan in each case as it presents itself; he must have a ready *coup d'œil*, so as to do the right thing at the right time and place, for what is excellent one day may be very injudicious the next. The plans of a great captain seem like inspirations, so rapid are the operations of the mind from which they proceed; notwithstanding this, every thing is taken into account and weighed; each circumstance is appreciated and properly estimated; objects which escape entirely the observation of ordinary minds, may to him seem so important as to become the principal means of inducing him to pursue a particular course. As a necessary consequence, a deliberative council is a poor director of the operations of a campaign. As another consequence, no mere theorizer can be a great general.

CHAPTER IV.

BATTLES.

Battles are the most conspicuous features of a campaign. A general may often put off a battle if he is not willing to fight; but a prolonged delay may become more damaging than a battle, on account of the many small combats and daily skirmishes which lead to no decisive result, and in the end occasion considerable losses; besides, the country around, from which the armies draw their supplies, may become exhausted. It is sometimes impossible to pursue a temporizing policy, because the country may not be of a suitable character, or the enemy may press so vigorously as to compel either a battle or a retreat. Again, a commander may fight a battle, when impelled thereto not by any necessity, but by the hope of great results from a victory. This he may do, if decidedly superior in the numbers and quality of his troops, when the enemy is expecting considerable reenforcements, or his different corps are too distant and disunited for mutual support, or his generals are on bad terms, or give evidence of irresolution and incapacity, &c. Although success is always doubtful, as fortune has quite as much to do in deciding battles as the talents and foresight of the commander-in-chief, there are other motives than those already mentioned which may induce him to fight a decisive engagement; such as the wish to relieve a besieged city; to extricate himself from a position where supplies are failing; to open a campaign by a brilliant stroke, which will demoralize the opposing army and animate his own. If it were possible to avoid battles and ruin the enemy by other means less influenced by fortune, the most ordinary prudence should lead to the adoption of the latter. But a battle is generally the only means of putting an end to the many other ills of a protracted war, and gives the wisest and speediest path to lasting peace; for, after a great victory, the conqueror may dictate terms; and even the defeated party goes from a severely contested field without disgrace, and may reasonably expect the consideration always due to courage and devotion to duty.

No general should determine to bring about so solemn and momentous a transaction as a great battle without having upon his side as many elements of success as possible. Therefore, all troops within a reasonable distance of his army, which are not absolutely necessary to guard other important points, should be called to him, for one battalion more or less may decide the fate of the day. If he has a great number of troops, only enough to engage the enemy thoroughly should be at first brought into action, the remainder being manœuvred against the flanks, or held in reserve, for use as occasion may require. The army should, if possible, be placed in such a position before the battle is begun, as to make the victory, if gained, lead to great results; as, for example, to the separation of the enemy from his dépôts and re-enforcements; to his being forced back, after defeat, upon some obstacle, like a lake, or river, or defile, where his forces may be wholly or in part captured or dispersed. When the features of the country are unfavorable to either army, it should fight with so much the greater order and obstinacy, that it may wrest from the enemy the advantages he has. When victory is essential for the preservation of life or liberty, men will perform prodigies of valor. A skilful commander knows how to bring success out of what inspires ordinary minds with despair. Cortez, after landing on the sea-shore the small force which was to conquer Mexico, burned his ships, and thus compelled his men to rely entirely on their swords, as there was no way left for retreat.

Art. I.—Definitions.—Orders of Battle.

If an army were always drawn up in the same manner, it would be certainly beaten by another whose dispositions were changed to suit varying circumstances. There is no invariable order of battle. It will always depend upon the locality. In a contracted space, an army will be drawn up in several successive lines; if, on the contrary, the front to be occupied is great, but a single line may be used. This use of one or more lines will depend on the topography of the battle-field, the circumstances of the moment, the forces and position of the enemy, the kind of troops forming the mass of his army, the character of his commanders, &c.

As a rule, troops in battle should be drawn up in several lines, as far as possible, to prevent the whole from being exposed to the first shock, and to enable one portion to be used to relieve or support another when necessary.

Usually, an army is formed in two lines, affording mutual support. The infantry and cavalry are distributed in these two lines, as the commander may deem expedient; but the artillery is always in the first line, and even in front of it, in order to have a wide field of view, and to take the masses of the enemy obliquely as they advance to the attack. The line of caissons is in rear of their pieces, but near enough to supply them readily. The first line of the order of battle is disposed in such a way as to be able to come into action as soon as the contending forces are within range of the small arms. Up to that moment the time is occupied in manœuvres, the formation of lines and columns, movements to the front and rear, &c. The second line keeps out of fire, taking advantage of undulations of ground, until its support becomes necessary.

Besides these two lines, a reserve is organized of all

arms of the service, which is held at some suitable point in rear, until called upon to move to some point where it is needed. This reserve, composed chiefly of cavalry and artillery, is under the immediate orders of the commander-in-chief, who uses it at opportune moments. With this reserve, and especially with the artillery, he re-enforces certain parts of his order of battle during the action; with the reserve cavalry, enveloping movements of the enemy are repelled by counter-attacks in flank; with the entire force of the reserve he strikes the blow which will decide the victory, or else repairs a disaster and re-establishes his line. This is of great importance, for there is danger in these movements of the original lines, which are made under fire, and, being not understood by the men, give them unnecessary alarm. More than one battle has been lost by an attempt to make a change in the order of battle during the actual progress of the engagement. As the movement of any part of the original line is, therefore, to be avoided, and a change of front is sometimes necessary, for this purpose, if for no other, a reserve is indispensable. The Romans adopted the rule of multiplying reserve corps behind the main army rather than prolong the wings, even if their force was much superior. This is a good rule still. In case of decided inferiority of numbers, the reserve must, of necessity, be small comparatively, because an army must not allow itself to be outflanked; *but there must be a reserve.* Upon the proper use of this reserve success is dependent, no less than upon preliminary arrangements and the courage of the troops. Generally, that army whose reserve comes last into action will be successful. A general shows his skill in compelling his adversary to employ his whole force while there is still unused, on his own side, the weight which is at last to turn the scales in his favor.

The battalions of the first line should be deployed, in order that the artillery may injure them as little as possible, and that they may deliver their own fire if opportunity offers; those of the second line are held in compact columns, so as to be readily moved to the front through the intervals of the first line, and against the enemy, or to execute any other necessary movement; or, finally, to leave openings through which the battalions of the first line may pass if driven back. The battalions of the second line will, therefore, be drawn up in close columns, as a rule; but this supposes the ground to be of such a character as to afford cover to these masses, otherwise the artillery would be very destructive, unless the line were held so far to the rear as to be beyond good supporting distance of the first. In that case, the second line would be deployed like the first, and at a distance of 300 yards or more. If the ground favors, it should be brought up nearer. The topography of the field has, therefore, a very decided influence upon this first element of the order of battle.

The troops of an army corps should be divided between the two lines. For example, a corps of four divisions would usually have two in each line. The two lines, being under the orders of the same chief, have a common interest, and will afford mutual support and assistance. If, on the contrary, entire corps are formed in each line, it sometimes happens that the commanders are jealous of each other, and do not act in concert; or the troops of the second line may give a cold support to the first; and the latter, expecting such a state of affairs, lose confidence and fight feebly. This arrangement is, however, quite frequently resorted to, and really has its advantages, for the commander has not his attention divided between two lines; and the second, being not so intimately connected with the first, is more disposable for such movements as may be necessary. In a broken country, the first is undoubtedly the best arrangement, inasmuch as it reduces the front by half, and lends itself better to varying ground.

The reserve should be composed of independent and entire bodies, in order to have independence and capability of rapid movement wherever its presence may be required.

A corps of sixteen battalions, divided into four brigades, two in the first line and two in the second, and with twenty-four pieces of artillery, would require about 1,500 or 2,000 yards front, or about one mile. This would be increased or diminished, according to the character of the ground. If in one line, the extent of front required would be about two miles, —a greater space than one man can readily keep well under his eye, especially in a wooded or rolling country.

Tacticians give different names to the order of battle, according to the manner of arranging it. It may be the *continuous* order or the order *with intervals.* With respect to the hostile army, it may be *parallel* or *oblique.* These different varieties are usually all found in the same army when engaged, some parts of the line being continuous and others with intervals, some parts parallel and others oblique, in some places single and in others double. But each of these dispositions must be examined separately in order to appreciate its properties.

The parallel order, as its name imports, is that in which the two armies are drawn in front of each other, so as to become engaged simultaneously along the whole line. Such must have been the style of the earliest battles, where no science was required, and success depended upon little else than individual strength, and upon the courage and obstinacy of the combatants. In this order, an equal effort is made along the whole front. If a victory is obtained, it is doubtless a complete success; but, on the other hand, a defeat is alike decisive. With equal bravery on each side, the victory belongs in advance to the more numerous party.

The oblique order, on the contrary, is a disposition by which a portion of the troops are carried against the enemy's line, while the remainder are kept disengaged. To use the technical terms, one wing is *thrown forward*, while the other is *refused*. The first is strengthened in every possible way, while the other is reduced as much as it can be, without compromising its safety. An attempt is thus made to outflank a wing of the opposing army, to crush the line at one point, while attention is drawn to others. If the enemy can be kept in a state of uncertainty, up to the last moment, by false attacks, and by a skilful direction given to the columns of attack; if they are promptly deployed, and only at the moment when the action is about to begin, the enemy will have no time for counter-manœuvres, or sending sufficient supports to the wing attacked, and will probably be defeated. Art may thus supply the want of numbers; and a small army, well commanded, may defeat a large one, whose chief has not known how, or has not been able, to mass his forces at the decisive point. This art is in our day much more important than with the ancients, because our lines are much more extended. The troops drawn from parts not attacked, to support those that are, sometimes cannot come up soon enough to check the confusion; and the battalions are overthrown one after another by the attacking wing of the army adopting the oblique order, which becomes more threatening by each successive step and by the deployment of its whole strength. At Leuthen this was the result, where Frederick, after threatening for some time the right wing of the Austrians, in order to induce them to draw a large part of their forces to that point, attacked their left with his best troops, took it in flank, drove it back upon the centre, and followed up so closely that the different corps, which came up from the opposite extremity to arrest his progress, were beaten in succession.

Among the generals of antiquity, Epaminondas seems to have specially apprehended the advantages of the oblique order, and to it he owed the famous victories of Levetia and Mantinea.

By refusing one wing, not only is the advantage secured of strengthening the other for the attack, but at the same time means are held in hand for recovering from the effect of a check, if one occurs, or for protecting the retreat, if necessary. In fact, the further the repulsed wing falls back, the more strongly it is sustained by the troops left to the rear, which come now successively into action. The front increases and becomes more formidable; a fresh attempt may be made, or, if the battle is certainly lost, a retreat may be commenced under the protection of the troops that have been but slightly engaged, and have not suffered. When the oblique order is properly employed, it gives many chances of success, as has been explained, and provides, as far as possible, against mishaps.

When a wing is refused, artillery of the heaviest calibre should be placed upon it, in order to keep the opposite wing of the enemy at a respectful distance, and to prevent the order from becoming parallel. The retired wing should also, as far as practicable, be in a good position for defence, which may, in case of need, be a set-off to its want of numbers. All natural features should be taken advantage of, such as woods, ravines, hills, &c., to afford support or cover to this wing.

The oblique order, properly so called, is undoubtedly preferable to the parallel order; but the latter has also its advantages, especially when it is re-enforced on one of its wings. It is very suitable for an army having a numerical superiority, that can present equal forces to the enemy's along the whole front, while at the same time it doubles its lines and effects a concentration of strength on one wing. In thus attacking the enemy in front, the victory will be more decisive, because the battle will have raged seriously at every point. This method, called the *parallel order re-enforced*, conforms to principles, because a principal mass is brought against one point of the hostile line, while at other points the opposing forces are equal. Thus, in fact, the parallel order re-enforced is analogous to the oblique order, and possesses the same properties, though not in so great a degree. In both there is a decisive effort at one wing, and simple demonstrations, or a less serious engagement, at the other.

If, now, we compare the continued order with that having intervals, either oblique or parallel, we will see that the first can only be used in large plains free from obstacles, where cavalry may always be placed on the wings. The other may be used in all cases of ground, and permits the use of the cavalry wherever it can act, whether at the centre or on the wings. If, in the continued order, the cavalry is put in the centre, it loses all its advantages; being kept down to the pace of the infantry with which it is in line, it loses its mobility, its chief, or, more accurately, its sole source of success; exposed, without possibility of reply, to the fire of the artillery, it must either stand still and be cut to pieces, or retire to a place of safety and leave a large gap in the line of battle. The chief of the cavalry would doubtless prefer to move forward, being supposed to be a man of spirit; but in charging infantry, which is in good order, his force will be overwhelmed by the fire, and probably dispersed; either way, the gap still remains in the line of battle.

This took place at the battle of Hockstaedt, which was gained, in 1704, by Eugene and Marlborough. At the battle of Minden, Duke Ferdinand, seeing Marshal de Contades place his cavalry in the centre, where the ground was favorable to its action, directed a portion of his infantry against it, giving orders that when the infantry had dispersed the cavalry, it should turn to the right and left and fall upon the flanks of the enemy's line, while an attack was made in front by the remainder of the army. This manœuvre, imitated from that at Hockstaedt, was equally successful. In the continued order, the cavalry must always be placed on the wings, that it may have independence and mobility; and if the ground there is unfavorable for its action, it is nearly useless. Such are the grave inconveniences of the continued order.

If, on the contrary, the different bodies of infantry leave intervals between them, the cavalry and artillery may easily act at opportune moments and wherever necessary. But in order that these intervals may not have the disadvantage, pointed out above, of leaving gaps through which the enemy may penetrate, the line should be formed in echelons, as shown in fig. 14. By this arrangement the cavalry may be held behind the echelons, ready to charge through the intervals, as shown by the direction of the dotted line A B. But this is not the sole advantage. The different bodies, not being attached to each other, can be more readily placed to suit the ground; have greater mobility in following up a success or withdrawing in case of reverse; the defeat of one does not necessarily involve those adjacent; one may rally under the protection of another. Notwithstanding the intervals, the entire line occupies the same front parallel to the enemy as if drawn up in the continued order, while permitting a still greater extension if circumstances require it. Each echelon flanks and supports the preceding one, so that if the enemy attempts to pass through an interval, he is taken in flank by the next echelon to the rear, while at the same time he is charged in front by the cavalry. Thus the order in echelons facilitates the application of the principle, that the different arms should be made to act on the ground best suited to their manœuvres; but at the same time the danger is avoided of leaving gaps in the line of battle.

The first echelon, C D, may be called the *point*. The second line and the reserves support the echelons of the first line, by adopting a similar formation or by accumulating near the *point*, or in any other way circumstances may demand or the general commanding may desire.

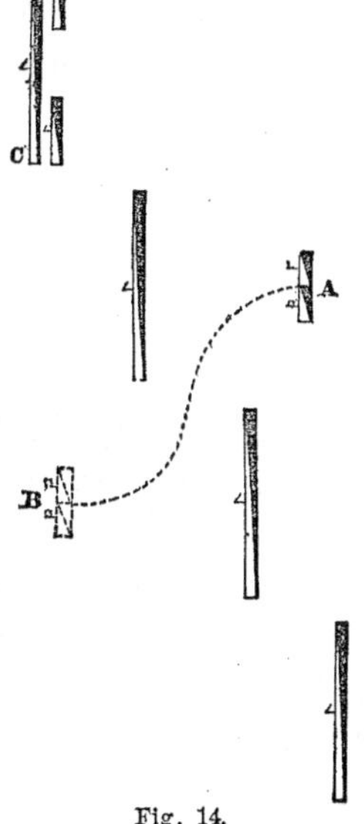

Fig. 14.

The *checker* order is the parallel order with intervals, the several subdivisions being drawn up in two lines, so that those of the second line are opposite the intervals of the first. This order is wanting in solidity, because the subdivisions of the front line are too far apart; on the same front it gives but half as much fire as a full line; with equal forces, it occupies double as much ground, which is almost always an inconvenience rather than an advantage, especially if producing diminished solidity. The artillery cannot be placed in the intervals of the first line, because it would attract towards the battalions of the second all the missiles of the enemy; for it is an established fact that artillery replies to artillery; this being the case, troops should not be in position behind batteries, as thereby a double mark is offered to the enemy. The intervals in the checker order or the quincunx order are the more dangerous in proportion to the proximity of the enemy, for he may, through such openings, penetrate in mass, overthrow the simple line which resists him, and drive it back upon the second line, where the same confusion may be produced. Generally, there should be no other intervals in a line of battle than those strictly necessary for the movements of the bodies of troops composing it.

This order, however, will be less vicious if the second line is formed of cavalry, because these troops may then charge through the openings before them. Thus the checker order, somewhat modified, might be partially adopted in an order of battle; for example, in a retired wing that must be extended to make a greater show. In a retreat, this order is employed, but rather as a manœuvre than an order of battle; it is then used to withdraw successively the different bodies of troops engaged with the enemy. When in a line of battle several of the subdivisions are drawn back in order to guard against flank attack, there is formed what is called a *crochet to the rear*. If the same subdivisions are thrown forward in advance of the line of battle, but so that they are attached to it, making an angle, there is formed the *crochet to the front.* The former may become the wedge and the latter the inverted wedge, when the two branches of the line are about of equal length. The wedge is essentially offensive, as an army would only adopt this formation with the intention of making an attack to pierce the opposing line, whilst the crochet to the rear would be adopted as a defence against an enveloping attack. The same may be said of the crochet to the front and the inverted wedge, but in a contrary sense; as the inverted wedge would be used when the object was to yield in the centre to the enemy advancing on that point, while enveloping his wings, it is a defensive formation, the crochet to the front being offensive. History describes a famous example where two armies were formed, one like the wedge and the other like the inverted wedge. It was the battle of Casilin, fought in 533 between the Franks and Romans, near Capua. The Franks were drawn up between two woods; they re-enforced their centre, forming a real wedge. The Romans were less numerous, and adopted an order having less depth but more extent, forming an inverted wedge. They gave ground in the centre and made their principal efforts against the wings of the Franks, the cavalry, in the mean time, attacking them in the rear after passing around one of the woods. The front of the Frank wedge, continuing to press forward, reached the Roman camp, the confusion of the pillage leading to their own defeat. The victory of the Romans was complete.

Art. II.—Employment of the Different Arms.

When the columns have reached the field of battle, and each is in the place marked out for it, the commanding general gives the signal for the action to begin, and each of his subordinate commanders prepares to execute the general instructions he has received, retaining a great latitude in the means of carrying them out. They know, for example, that the right is to make the great effort; that the left will be but feebly engaged, and in a manner only to give the enemy uneasiness and to retain him in his position; they regulate themselves accordingly, directing all their movements and manœuvres towards the common end. It is a sacred duty for each of them to give any support his colleagues may require. As the general-in-chief cannot be everywhere, it is for them to supply his absence, and make every arrangement that critical circumstances may require. They will do well if in nothing they act in opposition to his general plan. The greater their responsibility, the greater will be their part of the glory if success crowns their efforts, or even if deserted by fortune they fail, provided they have borne themselves bravely and honorably.

The battle is begun by skirmishers, who are thrown

out to the front, sometimes a thousand or fifteen hundred paces. Supported at suitable intervals by more compact bodies of troops and by light batteries, they attempt to check the enemy until the columns have had time to deploy; or else, when an attack is to be made, they cover the advancing columns, mask the dispositions for the offensive, repel the opposing skirmishers, observe the latest positions of the enemy, discover the vulnerable parts of his line, at the same time keeping him uncertain of the real point of attack, which should not be disclosed until the complete deployment of the columns and the onset is just about to be made.

Now the skirmishers should retire, either to permit the columns to become engaged, or because they can no longer resist the enemy and are obliged to give way, disputing the ground foot by foot. When within a short distance of the line of battle deployed behind them, they fall back rapidly through the intervals of the battalions and join their respective corps or rally in rear.

As soon as the front of the army is unmasked, the firing of the first line begins, the artillery first, and then the infantry, when the armies are sufficiently near for small arms to be effective. The firing is continued until some disorder is observed in the opposing ranks. This is the moment for forming columns of attack and charging. If the enemy retires before this offensive movement, the pursuit must not be headlong or disorderly, but the columns should be halted, perfect order restored as rapidly as possible, and then the onward movement resumed against the second line, which is usually disordered by the confusion of the first; care must also be taken that no part of the line of battle be advanced far beyond the remainder. It is only when the victory is certain, that the enemy may be rapidly followed up, and then with the object of making his retreat a rout.

When, on the other hand, it becomes necessary to fall back, the ground should be disputed foot by foot, order being preserved in the ranks as far as possible, that advantage may be taken of the least fault committed by the enemy. In every case, the second line will move up to support the first, and take its place, making, if possible, a counter-attack upon the advancing enemy. This will restore confidence to the troops, and prevent their thoughts dwelling upon retreat; and will, moreover, arrest the enemy or drive him back. The same ground is thus often taken and retaken several times by brave troops.

The cavalry may charge at any favorable time, and should never let slip any opportunity of rushing upon disordered infantry, or an exposed flank, or a badly supported battery, or of making a counter-movement against cavalry in motion. For this reason it is necessary to have convenient intervals, which will permit the cavalry to pass through, and to seize proper moments for falling with lightning speed upon the ranks of the enemy. Generally it is advantageous not to employ cavalry until a late period in the battle, as its effect is better then, and it is also in better condition to pursue the enemy in case of success, or to check his advance in case of reverse.

The artillery continues firing during the whole engagement. The cannon are often heard during the first skirmishing, and also give the enemy a parting salute when the victory is gained. The caissons should, therefore, be abundantly supplied with ammunition. The artillery is distributed in strong batteries, so as to produce large gaps at certain points of the line in front, into which the cavalry may rush. Without, perhaps, doing more absolute injury in this way than if distributed uniformly along the front, the moral effect is much greater. The soldier is filled with horror when he sees his comrades fall rapidly around him, and whole battalions swept away in an instant; he loses his presence of mind, he recoils; and if the cavalry presents itself at this moment, he is not in a condition for resistance. All the artillery should not be accumulated at one point, even if it were possible. It is easier to cover, by undulations of the ground, several separate batteries than a single one, containing all the pieces. The fire should, as far as possible, be convergent upon the important points of the line of the enemy, as, for example, a wing that is to be crushed, a salient angle that is to be enveloped, or a battery that is to be dismounted, &c.

The light artillery moves to the front to begin the action, following oblique directions, as far as practicable, both to unmask the other troops and to get slanting fires upon the enemy's lines. This kind of artillery should be bold, as its lightness enables it to escape with ease. Usually accompanied by supports of cavalry, it falls suddenly upon the flank of the enemy, advances, retires, halts, in order to avoid the attempts made against it, takes advantage of every favorable position to fire a few rounds. As soon as the victory is gained, the light artillery moves rapidly to the front with the cavalry, in order to break up the arrangements for retreat, following closely upon the heels of the enemy, giving him no rest, and in every way increasing the confusion in his ranks. It is of the first importance that light-artillery horses be of the best for strength and swiftness, and the cannoneers should be picked men.

The heavy artillery is collected in several strong batteries, either in the line in advance of the intervals between the divisions, or upon the wings; anywhere, in fact, where it interferes least with the infantry, and where the ground suits it. The smallest undulations are taken advantage of, to cover the pieces and give them elevation. A hill thirty or forty feet high gives an excellent position for artillery, as the fire is grazing and not too plunging, and the infantry supports can be kept near at hand and out of sight and fire.

Salient points of the order of battle are good positions for the artillery. Batteries thus placed give a concave formation to the first line, which is very favorable, as the enemy is enveloped and liable to a converging fire. In order, however, that a battery may be established on a salient point in this manner, it should be protected by some obstacle, either natural or artificial, from the flank fire it might otherwise be exposed to. Take, for example, the case shown in figure 15, where, in the centre, two batteries are protected from enfilading fire by the hill between them; on the left, another battery is masked by a village; and on the right, still another by a wood. The infantry is formed in line of battle between the batteries, and the light troops occupy the village, wood and hill. If no obstacle exists to cover the artillery and permit it to fire obliquely, as in the example shown, it is placed in the line, or, more exactly, a little in front of it, fires directly to the front, and ordinarily in reply to the artillery of the enemy. If the artillery is silenced, it fires upon the infantry, to break the ranks, cut up the columns, prevent their deployments and to keep

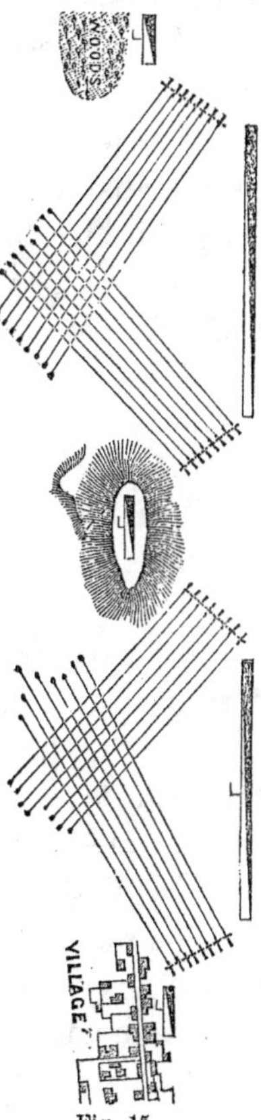

Fig. 15.

them at a distance; shells are thrown among the cavalry, which cause both destruction and alarm.

Firing at too long range should be avoided, for it wastes ammunition. The small effect it produces is discouraging to one party, and tends to embolden the other. The ravages of artillery are greatest within the range of grape. It would, however, be foolish to permit an enemy to approach so near undisturbed, because the effects of artillery are very considerable at fifteen hundred paces or more. The guns of smallest calibre are most advanced, while the pieces of heavy metal are placed at greater distances, as with a retired wing, upon heights, &c.

Art. III.—Offensive Battles.

The Reconnoissance.—The general arrangements for an attack can only be made after an examination of the position and force of the enemy. This examination is made by the commander-in-chief himself, and if the battle is put off to the next day, he must then, before giving his final orders, be certain that no change has been made by the enemy in his position. Without this precaution, the measures adopted might prove very unsuitable to the new state of affairs. Whilst he is on some elevated point, glass in hand, glancing rapidly over the field and the position of the enemy, the commanders of corps will be around him; he communicates his intentions, and gives notice of any modifications of previous orders, or reiterates them.

But if the battle is to be fought the same day, the skirmishers of the enemy must be driven in before the general can approach near enough to get a good idea of the advantages and inconveniences of the ground, on which the two armies are about to engage. He must at any cost reach some high point, where he can see, at least to a considerable extent, the surrounding ground, as well as the forces and dispositions of the enemy. While he is making his examination, and settling his plan of attack in accordance with what he has seen, he transmits to the different columns the orders relating to them; he sends the officers of his staff to guide them, and goes in person to that corps which will be first engaged, or is to take the principal part in the action; he ascertains from a brief conversation with the commander of this column whether his orders have been understood; he gives whatever explanations may be necessary, and hastens to other points. It is evident, from what has been said, that if the general is not with the advanced guard when the enemy is reported in presence, he should join it as rapidly as possible, in order to have time for the reconnoissance, and to make his dispositions for battle.

It is from this rapid examination that he determines the point upon which the strongest attack should be made, and whose possession will decide the fate of the battle.

Determination of the point of attack.—If the topography of the field is alone considered, it may be said, that generally, a height, a village, a wood occupied by the enemy, are so many points to be taken from him. In fact, a height is often the key of the battle-field, and should be the point upon which the first and great attack should be made, because such a point may give a commanding view and fire over all the surrounding ground. At the same time it serves as a screen, behind which new dispositions for attack may be arranged, and from the high ground a rush may at any moment be made upon the enemy below. Frederick said: "Always attack the mountain or highest ground occupied by the enemy, for if you force that position, all other points will fall into your possession; troops are always more vigorous and orderly in the first stages of an engagement. Do not, therefore, waste time and blood upon points of minor importance, and afterwards proceed with disordered and decimated battalions to attack the principal points of the enemy's position, where the greater part of his force will by this time be massed by the course of events if not by design."

The possession of a large village, with store-houses, is always a great advantage on the field of battle, because it may be very easily put in a defensive state, and is a kind of fortification. Such a village often becomes the scene of the most severe fighting, as, for example, Gross-Aspern and Essling, at the battle of Essling, and Castel Ceriolo, at Marengo.

A wood presents similar advantages. It covers and

masks the artillery; it is easily held by skirmishers; cavalry cannot approach it. A wood protects the flank of a line which rests on it, and the position of the latter cannot be held until the former is carried.

But the topography of the field of battle is by no means the only thing to be taken into account in determining the point of attack. There are considerations of another kind, much more extensive in their relations. The commander must first observe the distances separating the various hostile corps, for it will be proper to direct his principal efforts against the centre or one extremity, according as they are too much spread out, or are well connected with each other. He ought then to examine what is the position of the opposing army with reference to its line of operations, or any obstacles the country may present, in order to try to throw it back upon those obstacles or to cut its line of communications. The various reasons influencing to a movement in one or the other direction are often quite in opposition, and make the determination of the point of attack a problem of great difficulty, frequently admitting of a correct solution only from considerations of a character likely to escape common minds, and appreciated solely by those of vast powers.

It is impossible to lay down rules upon a subject like this, which is a matter of inspiration and genius; but we may group under three heads the considerations to be taken into account in determining the point of attack: first, those of a *strategical* character, relating to combinations of the highest order; second, the *grand tactical*, determining the principal manœuvre of the battle; third, *the tactical*, which, resulting from the nature of the ground where the troops are to act, may influence the details and execution of the general movements. It will be profitable to develop these ideas a little, and to cite some examples to explain more definitely what is meant.

If, for example, the opposing army has its right nearer the frontier of its own country than the left, and the line of operations is exposed, the attack should be made on the right, because there is a prospect in this way of cutting its communications, and separating it from its base; if a victory is obtained, the results may be very great. This is the *strategical consideration*. If the enemy is near a lake, a river, a great marsh, a thick forest, or a difficult defile, the *grand tactical* consideration, which is dependent upon the locality and the configuration of the battle-field and its environs, would lead to an attack on the side from the obstacle, in order to drive the enemy back upon it. If there is a height on the field of battle, attack that first, because success there will lead to decisive results. If a salient point is presented in the order of battle of the enemy, direct the first efforts upon this, because it may be enveloped and crushed. If his line is too disjointed and extended, strike at the centre, pierce it, and then beat in detail the separated wings. There are other grand tactical considerations. Finally, when the country is more open and practicable on one wing than the other, the desire to have unity and promptitude of movement, which is the *tactical consideration*, invites to an attack upon that wing. This last is the least important consideration, because vigorous efforts will enable troops to surmount almost any difficulties the ground may present, and often these obstacles once passed are a guarantee of further success, because the enemy supposed himself secure in that direction and was less on his guard. The tactical consideration should yield to the other two when there is antagonism among them in fixing upon the point of attack. When they are all in accord, there can be no doubt about the true point of attack, and none but the most unskilful commander would fail to detect it. Suppose an army, A B (fig. 16), drawn up in such a line as to

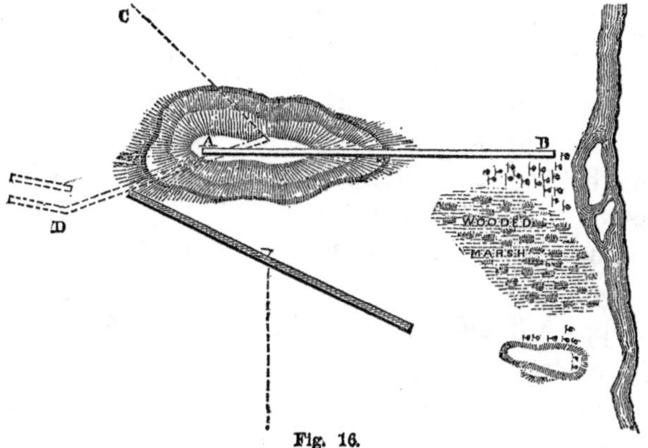

Fig. 16.

make an obtuse angle with its line of operations, A C, which is towards its right and rear; let the left of the army rest on a river, and the right be established on the high ground, A; let the ground in front of the left wing be a marshy wood, and that on the right open and unobstructed. Every consideration then leads to making the principal attack at A, for that is undoubtedly the key of the battle-field, and must be attacked, whatever may be the enemy's means of resistance there. If this height is carried, the line of operations, A C, is cut, and the enemy driven back upon the river. The army, A B, instead of taking the position supposed, should have formed the line A D, re-enforced on the wing D. In this position the assailant is very much embarrassed as to the choice of a point of attack. In the first place, he cannot attack the left, which is on the strong ground, and threaten the line of communications of the other army, without running the risk of being taken in flank, or turned by the re-enforced wing D, which may force him back upon the river. Or else, to avoid such a mishap, he will attack the right, D, at the same time holding the left in check; but then, even a success can have no decisive result, inasmuch as it does not lead to the possession of the commanding ground.

Suppose, in the next place, that the army, A B (figure 17), rests upon a lake in an open country, and that its line of operations, C D, is directed towards the banks, and to the rear of the lake. The strategical considerations would lead to an attack upon the left wing, B, with a view of cutting the line C D, but the grand tactical considerations would

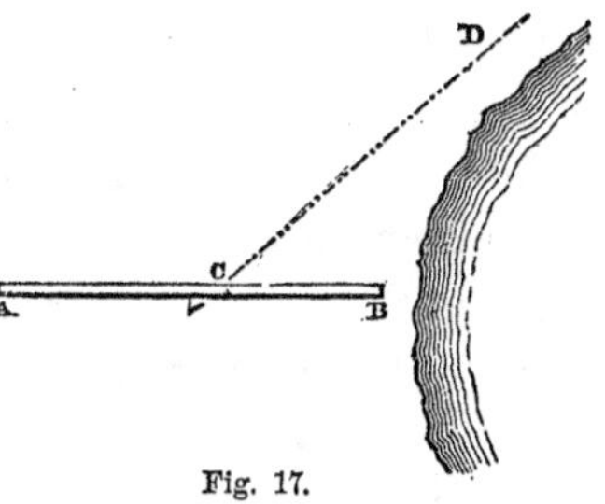

Fig. 17.

induce to attack the right wing, with the intention of driving the whole army back upon the lake. Here strategy and grand tactics advise opposite plans, and, in such a case, a battle should only be fought when the victory is, from other causes, almost certain. It would be much better to attempt by manœuvres to prevail upon the army, A B, to leave its position. In fact, an indecisive success on the right wing, A, does not drive the enemy upon the lake, nor cut his line of communications. On the other hand, it will not do to violate the law of grand tactics and attack at B, for, if the attempt failed, the enemy might take the initiative and drive the attacking army into the lake. If a battle must be fought, the strategical considerations would, in this case, be deemed less important than the tactical, although the general rule is otherwise. In fighting a battle, we must always take into account the contingency of a defeat, and not place an army in such a position that defeat would be destruction.

If an army is posted between two rivers, covering

its flanks, and the line of operations is perpendicular to the line it occupies at the middle point, there is nothing special to be gained by attacking on one side rather than the other—it is entirely a tactical question.

When the point of attack is once chosen, it should be pressed vigorously, all the arms of the service co-operating, and as many troops being employed there as can be without confusion. The enemy should be kept in a state of uncertainty up to the last moment. If the first attempt fails, renew it again and again, if necessary. Such pertinacity often leads to victory. If the first effort is successful, the enemy should be vigorously followed up, and no time or opportunity afforded him to recover himself.

Attack of heights.—Although it is a rule to attack heights in opening an engagement, it should only be done when they are attainable, and the troops, upon reaching the high ground, will have room to extend themselves; for, if such positions can only be approached through narrow defiles and under fire, and the troops must advance with a very small front, there is little prospect of success. Such a position must be turned. But it must be recollected that, in seeking to turn a position, the turning party runs the risk of being taken in flank and seeing his own communications cut. Such a movement should, therefore, only be attempted when there is no opportunity for the enemy to debouch on the flank.

High ground may not always present its front, but

may be a chain of hills, perpendicular to the general line of the two armies, both resting one wing upon them. In such a case, the attack should proceed from the higher to the lower portions. Thus, the two armies, A B and C D (figure 18), having each one wing in the plain and the other on the high ground, the general direction of which is perpendicular to their front, the army A B should attack by its right, in order to dislodge the left of the enemy, D, before bringing B into action. The centre, E, follows the right, and serves as a connection between the wings. The attack is thus progressive from right to left, and becomes general only when the right and centre have repulsed the left and centre of the enemy. The order of battle, from being at first parallel, becomes oblique and in echelon.

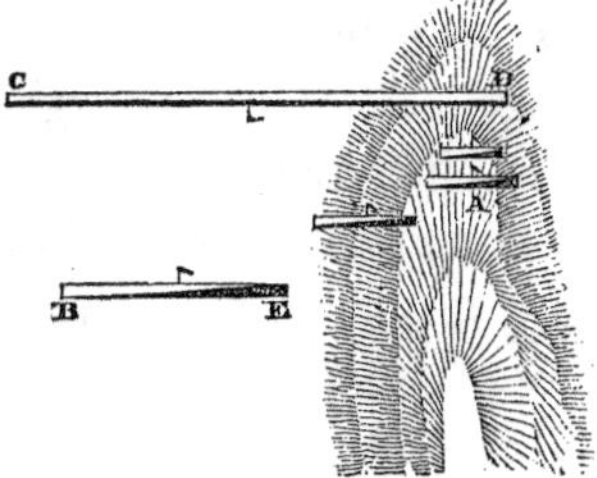

Fig. 18.

The enveloping attack.—If the attacking army is sufficiently numerous, the enemy may be enveloped upon a single wing, or upon both. If the attempt is successful, victory is nearly certain, as the long shallow lines of a modern army are almost defenceless against such an attack on the flank. A too extended movement must, however, be guarded against in such an operation, for a line with great development becomes proportionately weaker, and may easily be pierced, as the Russians discovered at Austerlitz. It is better to keep the army concentrated and to fight in front, than to attack a flank in the manner referred to. The enveloping attack requires, therefore, superior numbers, and precautions should be taken to keep the enemy occupied throughout his front while the enveloping mass assails the flank, but by no very extended movement. For still stronger reasons are those very great circuits to be avoided, by which a body of troops is sent to attack the enemy in rear. They are in opposition to a fundamental rule in war, which requires the concentration of all the disposable forces when an engagement is to take place.

The flank attack, properly so called, differs from that just referred to, since, in this case, the flank of the enemy is assailed not only by a single corps while the others hold him in check along the front, but the whole army is drawn up in a line crossing that of the enemy, at a greater or less angle. It is certainly to have gained an important advantage to have been enabled to take up such a position, since only a part of the troops of the enemy can be brought into action, and as the other portions come up they may be successively overthrown. The advantages of the oblique order are here seen in all their force. But is it probable the enemy will stand still and suffer himself to be thus taken in flank, if his attention is not drawn in some other direction? Is it not easy for him, by a similar operation, to counteract a circular movement such as must precede an attack of this kind? May he not even attack at some favorable moment, while the army is filing before him and in taking position preparatory to battle? All this may happen, and hence the flank attack is extremely hazardous. If the great Frederick owed most of his victories to this kind of an attack, it was because the opposing army manœuvred with excessive slowness and heaviness, and because he knew how to conceal his preparatory march by a demonstration in another direction. As a rule, the turning party is also turned; if he threatens the line of communications of the enemy, he exposes his own at the same time, so that a manœuvre-march, made with the view of massing the army on the flank or rear of the enemy, can only be justified by a favorable locality, and the possibility of its accomplishment without exposing the line of operations. For example, fig. 19, the army M may, without much risk, and with considerable prospect of success, manœuvre against the flank of the army N, if the line of operations, P Q, which it is obliged to leave temporarily, is protected by a river and by difficult ground on its flanks. A weak corps, *m'*, stationed on this line, behind a stream with a deep bed, making a strong defensive point, may hold the enemy in check long enough to allow the main body, M, to move around along the dotted line from *m'* to M. Even in this case, a body of troops, *m*, should be placed in such a position as to

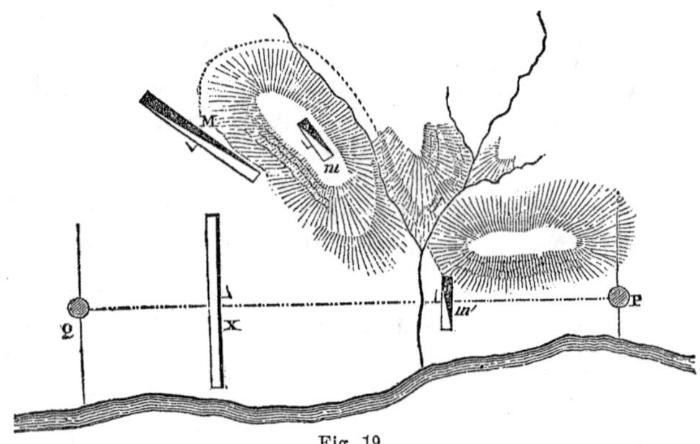

Fig. 19.

have a commanding view of the enemy, and give warning of any attempt on his part to strike at M in flank while on the march. Although the corps M, *m*, and *m'* are somewhat separated from each other, the connection is sufficient, as the intervals cannot be penetrated by the enemy. Napoleon practised a manœuvre almost identical with this, during which the famous battle of Arcola was fought.

It seems, therefore, evident, that if an attack in mass by the flank is sometimes practicable, it is only so under the cover of natural obstacles, and with certain precautions; such as concealing the movement, occupying in sufficient force all points which, if in possession of the enemy, would enable him to fall on the flank of the moving mass. Generally, on the battle-field, the enemy must be fairly attacked, without consuming time in movements to turn him, which often prove dangerous, and frequently useless, because easily avoided.

A body of troops may sometimes, by a simple deployment to the right or left, gain ground in that direction so as to outflank one wing of the enemy. Suppose, for example, it is desired to attack, with five battalions, an enemy in equal force, and to take him in flank. Two columns, A and B (fig. 20), will be

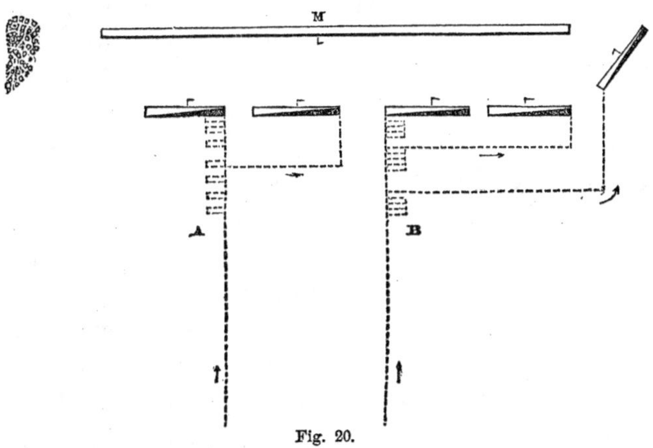

Fig. 20.

moved forward towards the centre of the hostile line. The first, A, will consist of two battalions at half-distance by divisions; the second, B, of three battalions closed in mass, and with one color concealed. The columns will seem to the enemy to be of equal size, and he will not know where the principal attack is to be made. He will think it is to be in front and parallel, until the deployments, under cover of the skirmishers, show him a fifth battalion on the extremity of his left wing. The skirmishers that should be displayed against his right will contribute to the success of this manœuvre. If the attack is vigorously made before the enemy has recovered from his astonishment, it will be almost certain of success. What a small corps of five battalions has done in this case, a large army with numerous columns may repeat. Every head of a column shown to the enemy increases his uncertainty, and this is a reason for pushing them forward equally, as if the intention was to adopt the parallel order of battle. Being in this state of uncertainty until the last moment, he cannot prepare well for the blow which is to fall upon him.

If the ground on one of the wings presents any feature, under cover of which a body of troops may slip along unperceived, advantage should be taken of it. Even a small body, appearing at an unexpected place and time, will astonish the enemy, and if the attack is pressed at this moment it will certainly succeed.

Attack upon the Centre.—Although the centre is the strongest part of an order of battle, there may be circumstances making it advisable to direct the main attack against that point. Suppose, for example, that a height, the key of the field, is in the centre of the line. In such a case the bull must be taken by the horns if a decisive result is desired, and consequently the most vigorous efforts should be made to carry the

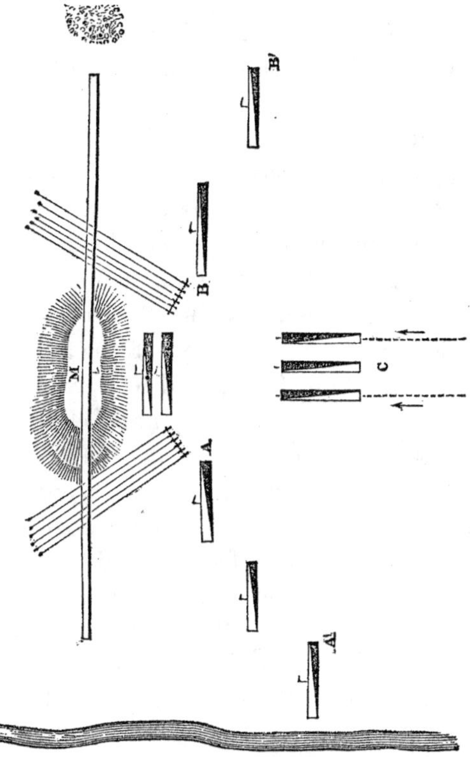

Fig. 21.

high ground, the possession of which is equivalent to a control of the field. An advantage obtained on one wing would amount to nothing so long as the enemy held the high ground, and it would be always necessary to drive him away from that part at last, or else retire from the contest. Therefore, however perilous this attack may be, there is no time for hesitation; the attempt must be made as the only possible chance for success, unless, indeed, the position can be turned, in which case it would be better to manœuvre than to fight. If the attack is made, the best troops should be selected for that purpose, and every possible pains taken to insure success. The natural order in such a case is the oblique order in echelon by both wings, presenting a figure somewhat similar to the wedge of the ancients. The point, A B, of this wedge (fig. 21), intended to pierce the centre of the army, M, is re-enforced by doubling its line and by drawing nearer the reserve C. On the other hand, the echelons, A′ and B′, of the two wings are formed in a single line, as they are to be but slightly engaged, only, indeed, to a sufficient extent to hold the opposing wings in check. The artillery, being non-effective against the high ground, is put in position to fire obliquely on those parts of the line adjacent to the centre, and thus in a measure to separate the centre and wings. In short, every effort must be made to have this attack successful. If not, the battle is lost, and it is time to think of retreating. The figure, showing the army M supported on the flanks by natural obstacles that cannot be turned, explains sufficiently why it has been necessary to attempt to carry the centre.

There is still another case when it is proper to attack the centre, even when the opposing army is well concentrated. It is when the latter is in front of a defile, which offers but few advantages for a retreat; because if the centre is pierced, and the mouth of the defile seized, the enemy is lost, being obliged to fly in confusion, abandon his baggage, and lose many prisoners. The entrance of the defile is in such a case the key of the field of battle. At Waterloo, Napoleon directed his main attack against the centre of the English army, because the latter had the great forest of Soignes behind it, and the only line of retreat for the cavalry and the baggage was along the road to Brussels. This attack failed only because the defenders of the position were as brave as the assailants, were more numerous, and towards the end of the battle received considerable re-enforcements, which decided the victory in their favor.

The attack on the centre, which is but an exceptional case when the opposing army is well concentrated, becomes the rule when the army is much spread out, and the different corps in it are too far apart. By making simple demonstrations upon the extremities of this long line, and pressing vigorously upon the centre, it will almost surely be pierced, and the two wings will never succeed in reuniting. It will be practicable under such circumstances, if time is used to good advantage, to envelope and destroy the wings in succession, or else oblige them to retreat upon divergent lines. In 1808, the Spanish army of 45,000 men took up a position in front of Tudela; but the general, Castaños, instead of assembling them on a front of two miles, as he should have done, extended them over three and a half. Marshal Lannes, who commanded the French army, perceived the weakness of such a disposition of troops, and attacked in the centre. One division of infantry promptly broke the line, and the cavalry, passing through the gap, and turning to the left, enveloped the Spanish right wing, and completely routed it. The left wing could make no further resistance, although composed of the best troops, and retired with precipitation.

Concentration.—In whatever direction the attack be made, whether upon one of the wings or the centre, whether the order of battle be parallel or oblique, the mass intended for the attack should be composed of all arms, and should contain as many battalions, squadrons, and batteries as can be brought together and can act without confusion. No effort should be spared to make the blow a decisive one. The lines should be doubled and the reserves brought near. Success depends upon the simultaneous, cordial, and vigorous action of all the individual parts.

To give an idea of the method of effecting such a concentration of force and action as is necessary, we will take the case of an army of four divisions, required to act in an open field, with the intention of making a powerful attack from its right, after engaging the centre. The army may consist of the following troops:

40 battalions	30,000
30 companies of riflemen or sharpshooters	3,000
12 squadrons	2,096
16 batteries, each of 6 pieces	2,800
4 companies of engineers	400
Total	38,296 men.

It will be observed that the proportion of cavalry and artillery in this army is small.* It is supposed that the advanced guard has rejoined the main army and that all the troops are with their respective divisions. The entire force of cavalry is joined to the division which is to form the reserve. All the disposable artillery and riflemen are also held with the reserve.

The principal attack is to be made by the right, and the right division is consequently strengthened by the addition of five companies of riflemen, giving it ten, each of the two other divisions five, and the reserve ten. Three divisions are to form the line of battle, each presenting a front of four battalions, and the remaining six to be disposed of in the best manner to carry out the plan of battle.

This arrangement is agreed upon the evening preceding the battle, but is only definitely settled the next

* Jomini says: "As a general rule, it may be stated that an army in an open country should contain cavalry to the amount of one-sixth its whole strength; in mountainous countries one-tenth will suffice."

Of artillery, the same authority says: "The proportions of artillery have varied in different wars. Usually three pieces to a thousand combatants are allowed, but this allowance will depend on circumstances."—TRANSLATOR.

morning, when an examination of the field has shown that the enemy has made no changes in his dispositions. The following order is then sent to the division commanders:

"The principal attack will be from the right, the left being retired.

"The army will advance in three columns, at sufficient intervals for deployment. They will regulate their motions by the right, and neither will get in advance. The reserve will follow the central column. Each column will be preceded by an advanced guard, composed of its riflemen, the flank companies of the leading brigade, and a battery.

"The deployment will be effected in the usual order, each brigade forming one line, the artillery on the right, the riflemen in the intervals of the battalions, and the flank companies on the wings. If at any time during the battle the artillery moves to the front at any point, the battalions will form columns, to make room for its passage."

Under this order the divisions are moved to the front, preserving their proper intervals, and preceded by their advanced guards some 1200 or 1500 paces. They commence to deploy at a signal of three guns fired by the reserve. When the advanced guards are sufficiently near the enemy, the riflemen are deployed as skirmishers to engage the skirmishers of the enemy. The flank companies, already referred to, form their supports, each keeping near the battalion to which it belongs. The artillery will, in the mean time, have opened upon the enemy whenever his masses have been discovered.

The combat of the skirmishers is continued until they are driven in or called in nearer to the line of battle, which is formed by this time. The batteries move up at a trot, to join the troops already in position. The supports relieve the riflemen, who rally behind them. One-half forms the line of skirmishers and the other the supports, and in this order they fall back gradually towards the intervals of the battalions. The batteries are first unmasked and immediately commence firing. The riflemen take position in the intervals of the battalions, and the flank companies, which have formed their supports, doubling on the wings.

The two right divisions now become engaged, opening fire if the enemy is in good range, or charging with the bayonet. The left division cannonades the right of the enemy and makes a display of troops; the only object there being to hold that part of the opposing line in position.

Figure 22 shows these arrangements, and the manner in which the generals of division have conformed to the spirit of their instructions. The general of the right division has withdrawn one battalion from each line to form a small reserve, which he has placed behind the centre. He will engage in the parallel order. The general of the central division has formed his

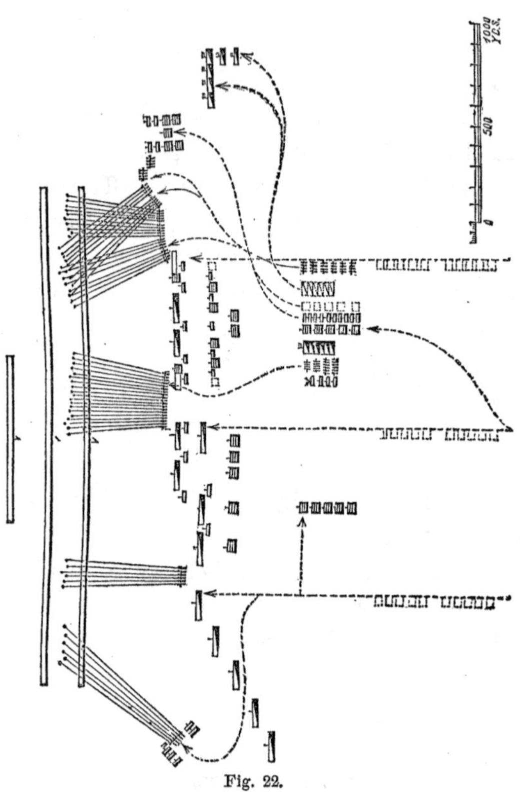

Fig. 22.

first line in two echelons, but has doubled the right battalion and placed all the riflemen with the first line. The second line has nothing peculiar about it, except that the three right battalions are nearer together than the two on the left. This is in conformity with the primary idea of making the main attack towards the right and refusing the left. The general of the left division, in order to retire the left, has formed his first line in echelons by battalions deployed at 100 yards distance from each other. One of his batteries has been thrown to the left and front, in order to cover his echelons and lead the enemy to expect an attack on that side. The five companies of riflemen form a support for that battery, and the flank companies of the battalion form the line of skirmishers. The second brigade remains in column to the rear.

The general-in-chief, having placed his reserve behind the centre of the right division and seeing the engagement commenced, judges that the moment has arrived for the development of his plan. He detaches one brigade of the reserve, with four companies of riflemen, six batteries, and half the cavalry. He sends verbally the following orders: "The artillery to move up to the right and left of the batteries of the right division, and to take the line of the enemy by as slant a fire as possible; the infantry to be massed on the right of the artillery, as supports, guarding against attacks of cavalry (the smoke of the cannon will facilitate this movement); the cavalry to follow the infantry, keeping to the right and rear, and being prepared to take in flank the cavalry of the enemy, should it attempt to attack." Two of the reserve batteries are also sent to re-enforce those of the central division. The figure (22) shows these movements executed.

The right and left battalions of the right division

form column, to make room for the artillery; and those of the second line close in towards the centre, to avoid being behind the batteries. The infantry brigade gains ground to the right, preceded by the riflemen, who are ready for deployment as skirmishers. This brigade may either charge in mass, deploy the leading battalions, or form squares to close the front line. The cavalry is deployed, to appear more numerous, two squadrons being always kept in column.

The order of battle, at first parallel, thus becomes oblique, and the enemy will be turned if he does not hasten to withdraw his left wing. In the figure (22) the two armies are four hundred or four hundred and fifty yards apart, which supposes that the enemy has brought into action as yet only skirmishers and artillery, or has already fallen back to avoid the flank attack with which he is threatened. But that has nothing to do with the movements just described, which are so much the more decisive if the armies are in close musket range.

The reserve being weakened by these movements, the general sends orders to the second brigade of the third division to draw near the centre, to act as a second reserve in case of need. When the first division is about to move to the front, its two deployed battalions will first advance, in order to give a good volume of fire within effective range; when the time for charging has arrived, these battalions form column, those of the second line take their places in the intervals, and all advance together. The muskets will not be brought to the position of "charge bayonets" until the columns are within ten or twenty paces of the enemy. It is a bad plan to bring them down sooner, as the marching is more awkward and the alignment lost. The enemy seldom waits to receive such an onset. If a second deployment is necessary, it will be done by the battalions of the second line, those of the first taking the places of the latter. During these manœuvres, the riflemen and flank companies, placed in the intervals of the columns, will have opportunities of using their fire-arms effectively.

In the general movement, the centre will be regulated by the right, constantly endeavoring to gain ground. If any gap occurs in the line on account of the manœuvres, it should be filled by the battalions of the second line. The officers should use every exertion to have the engagement progress from right to left, all disconnected movements being carefully avoided. Finally, the left, which has been retired in the beginning, becomes engaged, and the action is general. The troops along the entire line are engaged with the enemy, who, being outflanked on his left and assailed in front by doubled battalions, ought to be defeated, unless he is very superior in artillery, for this arm, in its present state of improvement, has very much to do in the decision of battles, especially in open ground.

Enough has been said to give an idea of the method

of concentrating upon a given point a large mass of troops, and of handling them in action. It will be observed that upon the right, now become really the centre by reason of the offensive movement which has been explained, the troops are really drawn up in three lines. This arrangement might have been made still stronger by forming a greater number of battalions in column and closing them more, but they would thus have afforded a better mark for the artillery of the enemy, and there would have been the additional inconvenience of having a portion of the troops reduced to a state of inaction. With modern weapons, the problem to be resolved is to cover as much front and to obtain as long a line of troops firing as possible, without having too much weakness in the line of battle.

In the example selected, the ground is supposed to be level and open, so that the artillery would be the most important arm. That is by no means the general case. The dispositions must always be made to conform to the ground. Where it is more broken, the troops will remain more in columns, which will move over the practicable parts of the field; the combat of skirmishers will be continued longer, and, as the artillery of the enemy is less effective, they may advance nearer; the bayonet will be more frequently used; the artillery will regulate its movements by those of the infantry, and will not be so much massed, on account of the want of suitable ground; finally, the cavalry will be little else than spectators until the pursuit begins.

Even in an open country, manœuvres cannot be conducted with such methodical regularity and exact distances as have been seen in the example given, if the enemy is met on the march and a battle ensues the same day. It may begin when but two divisions are at hand. Sometimes a single division, with the advanced guard, will be obliged to receive the first attack of the enemy, while the other divisions are coming up in succession into line. An oblique line in echelons will necessarily result under such circumstances, and the commanding general, having made provision for first emergencies, should give suitable orders to the different corps as they successively arrive on the field. Upon such occasions a skilful commander demonstrates his ability; he has no time for protracted reflection; his arrangements must be made upon the spur of the moment. His combinations should be simple, his orders brief, and he must be cool when every one around him is in motion and excitement. Any man of ordinary capacity and some experience in moving troops may, after reflection, make suitable dispositions of an army for battle, but he must be born a general who can improvise them under fire.

Pursuits.—It may happen that while a decisive attack is made on one wing, the enemy may be doing the same thing on the other, and the two armies may each be victorious on one wing and defeated at the other. It is, therefore, important to modify the ardor of the troops, and instead of indulging in a headlong pursuit, to keep them well in hand, and move them all or in part upon the flank or rear of the other wing of the enemy. It will be sufficient to follow up fugitives with the light cavalry, supported by light artillery and a few battalions of infantry. In the mean time, the wing of the enemy which has been victorious may be enveloped, and its defeat made so much the more complete as it had advanced beyond its original line. At the battle of Naseby, fought in 1645, between Charles I. and the parliamentary forces, Prince Rupert defeated the troops in his front and pursued them vigorously. But Cromwell, who had defeated the royalists immediately in his front, did not pursue, but turned on those that were still fighting, took them in flank, and entirely routed them. If Prince Rupert had pursued this course, he would probably have saved Charles his crown.

After a first success, therefore, the pursuit of the enemy should be made with circumspection; the ranks should be reformed, and order preserved in anticipation of further attacks. When it appears safe to move forward, the pursuit may be recommenced in such direction as to separate the corps of the enemy and prevent their reunion. While the battalions are preparing for this pursuit, the light troops will harass the retiring enemy; the artillery will follow up and give him no rest; the cavalry will charge vigorously those corps which seem to be recovering their order, and will cut them to pieces or oblige them to lay down their arms.

In following out the rule of reforming the ranks before pursuing actively the enemy when he is thrown into confusion at any point, care must be taken not to lose too much time in making the necessary rectifications of alignments. Perfect alignment is not so important to the preservation of order as the touch of man to man. Keep the troops well closed and push forward; nothing more is necessary, to overthrow completely an enemy who has once yielded ground. If, on the contrary, time is lost in aligning the ranks as upon a drill-ground, the enemy will be enabled to reform and offer battle a second time.

It appears, therefore, that two extremes are to be avoided when the enemy is falling back. Too much circumspection may prevent a first success from leading to decisive results, and too much rashness may cause total ruin. A union of prudence and boldness is necessary. A commander must look not only before him, but often to the sides and sometimes behind. He must know what is passing in the neighboring corps; must observe whether he is supported; if the line of which he forms a part is maintaining its position; if the reserves are engaged; if he must depend upon his own means, &c.

The commander-in-chief will direct the movements

of the whole army in such a way as to render his victory decisive, by gaining as much ground as possible in the direction of the line of retreat of the enemy, in order to cut him off; the last reserves will be disposed in such a manner as to overthrow any remaining resistance. While the different corps are pressing the defeated enemy, turning them, driving them upon obstacles, making prisoners and capturing material, the general should take some rest, dictate orders for a bivouac of the army, and for moving the next day; he should give attention to the wounded, and should signify his satisfaction to the troops in an order of the day, in which he will recount what each corps did in gaining the victory. Finally, he should take measures to replace expended munitions, fill up the ranks of his army, and draw from the country such supplies as he needs and it can furnish.

Art. IV.—Defensive Battles.

The weaker army is usually obliged to receive an attack from the stronger. It chooses as favorable a site as possible and awaits the enemy there, endeavoring thus to supply its numerical inferiority by superiority of position.

Positions.—A good position is one which, not too high, still overlooks the surrounding ground and affords space enough for deploying the troops. The ground should be sufficiently smooth and unbroken to facilitate the manœuvres of all the arms of the service, but especially the artillery and cavalry. Its extent should be proportioned to the strength of the army; and it may here be remarked, that in a good position the same troops may occupy a greater front than in a plain, as there is not the same necessity for doubling the line throughout. It will be sufficient if two lines are drawn up on the most easily accessible parts of the field; one line will answer at other points, and moreover considerable spaces may be left between different corps, which will be occupied by skirmishers. Some localities permit such deviation from general rules. All salient points being held by the artillery, or even by sharpshooters and infantry, the enemy cannot, with a prudent regard for his own safety, expose his flanks by penetrating between them and finding the reserves in his front at the same time. A division of sixteen battalions, four batteries, a battalion of sharpshooters, and four squadrons may conveniently occupy a front of two thousand yards.

The wings of the position should rest upon strong natural obstacles, such as large marshes, a lake, a deep river, a wood, impassable rocks, which secure the army against attack in flank, or oblige the enemy to make wide détours to turn the flank. In front the ground should fall away gently, so that it may be thoroughly swept by the artillery; and it also permits forward movements against the enemy when advisable. There should also be at intervals along the front clumps of trees, villages, farm-houses, or enclosures, which, being occupied in a proper manner, become formidable salient points, that furnish cross-fires and must be carried by the enemy before attacking the real line.

A position whose front is covered by a river, or by very steep and difficult ground, is only suitable for an army which is too weak to venture a forward movement against the enemy, but must be content to remain strictly on the defensive. Even in such cases there should always be efforts made to facilitate offensive returns upon the enemy at opportune moments, as these have a powerful effect in arresting an attack.

In rear the roads should be good, in order to facilitate a retreat in case of reverse. A single road is not enough for the easy and prompt evacuation of a battle-field. Moreover, if the enemy should seize that single road to the rear, the army is lost; this is especially dangerous if the road leads from one of the wings, and not from the centre. The most favorable case is that where there are several good roads through a wooded and broken country, where the army may find good positions for checking the enemy. An open country is dangerous, as nothing is more to be dreaded than charges of cavalry upon a retreating army which is in more or less confusion.

Besides the roads leading to the rear, whose general direction is perpendicular to the line of battle, it is well to have in rear of the line a cross-road from right to left, so that the artillery may move freely from point to point. The best place for this road is behind the second line. Besides the foregoing advantages which the position should offer to the defending army, the low ground that is left to the enemy should be cut up and obstructed by ditches, ponds, hedges, walls, &c.—obstacles that are not impassable, but greatly tend to delay and confuse the advancing troops, especially if under fire.

The general form of the position should be concave towards the front, if its extent is limited, but convex if there is considerable development. In the first case the army occupies the entire extent of the position, and has its wings strongly posted, so as to be in no danger of being enveloped; it has only to stand and deliver a close, converging fire upon the enemy. In the second case, however, as it will generally be necessary to move troops from one part of the field to another, to repel the various attacks of the enemy upon the extended front, it is best to have the position convex towards the front, as the troops may then move on the chords of the arcs passed over by the enemy. But, properly speaking, such a position as the last is not a good field for a defensive battle, which should always be in strict proportion to the army holding it; it should rather be deemed a piece of ground offering facilities for manœuvre-marches, and giving opportunities for prompt movements upon threatened points, or for effecting a concentration of troops at vulnerable parts of the enemy's position. The advantages of such ground are rather strategical than tactical.

Positions actually occupied, seldom possess all the advantageous conditions that have been mentioned, and the best are those in which the most of them are found. The general shows his skill in making the best use of every advantage presented by his position, and in supplying what is wanting by field-works, or a good disposition of his troops, or both.

In order that a village may form a strong point upon the front or flank of a position, it should be of solid construction. Wooden houses, far from being favorable in the defence, may become very hurtful, because they may be so readily fired. Villages may be occupied by several battalions, by sharpshooters and by artillery, according to their importance; the walls of the outer enclosures, by being loopholed, greatly assist in the defence, and guns may be concealed behind the houses in favorable positions for procuring a flank or slant fire upon the troops of the enemy. The line, of which the villages are the salient points, should be near enough to support them readily, and prevent their being turned and surrounded.

However excellent the features of a position may be, an army should seldom be satisfied with a purely passive defence. On the contrary, it should always take the offensive whenever a favorable opportunity offers for striking an effective blow. By moving out to attack, instead of waiting to be attacked, the weaker party, by seeming boldness, may conceal real weakness; by unexpected offensive movements a detached or venturesome corps of the enemy may be cut off; at any rate, he will be made more circumspect in his proceedings, and the *morale* of the defensive army will be elevated.

Disposition for the Defensive.—Since strong positions are not always to be found, and obstacles are not always at the proper distances apart to form good points for the flanks to rest upon, it is necessary for a general, while knowing how to take advantage of any favoring features of the ground, to be able also to supply the want of these, as far as practicable, by a judicious arrangement of the troops at his disposal. The essential thing is to make the wings secure against a flank attack. To guard these weak points against charges of cavalry, it is sufficient to place there several battalions, which may be drawn up in squares. At the battle of Molwitz, gained by Frederick in 1741, the cavalry of his right wing had been routed, and the infantry were about to be taken in flank, but the victorious cavalry of the enemy was checked by three battalions, which for want of space, could not be deployed, as was expected, in the line of battle, and were drawn up behind the wing. This infantry, although repeatedly charged, stood firm until Marshal Schwerin brought up the left wing, hitherto retired, and gained the victory. Had not these three battalions been accidentally in the position they occupied, Frederick would probably have been defeated. It is seldom an evil when want of space compels some of the battalions to remain in column behind the line, or several companies behind a battalion.

The wings may also be strengthened by forming behind them in echelon several squadrons of cavalry, to make counter-charges, in case the enemy should attack in flank. In the orders for the battle of Hohenfriedberg, Frederick directed a regiment of hussars to form a third line behind each of the wings of the army, "either to cover the flanks or to be used in the pursuit." If no cavalry can be formed for such a purpose, battalion squares, in echelon on the wings, will secure them against enveloping attack from the cavalry of the enemy.

The more serious efforts of infantry against the flanks during a battle, while the front is strongly engaged, may be paralyzed by causing the second line to outflank the first, and the third to outflank the second, so that the enemy is himself turned when he attempts to assail the flank, or he is obliged to make an extended and hazardous circular movement, should he endeavor to turn all the lines at once. In this way, during the first attack of the enemy, and until his real designs are displayed, as few troops as possible are exposed. One of the wings may thus be kept disengaged, or the whole disposable force brought into action, as occasion requires, by successive portions, without interference or confusion.

In order to give a clearer understanding of the preceding remarks, the same army of four divisions, already described as fighting an offensive battle, will now be supposed placed upon the defensive. A plain will be taken as the field of battle, entirely bare of obstacles, that we may examine simply the arrangement of the troops, and strip the example of every thing extraneous.

The commanding general, wishing to retain a strong reserve, withdraws two battalions of each division in order to form a supplementary brigade, which he attaches to the reserve division. The divisions of the line of battle have each eight battalions, two batteries, and four companies of sharpshooters. The cavalry and the remainder of the sharpshooters are joined to the reserve, which is thus composed of sixteen battalions, twelve squadrons, ten batteries, fifteen companies of sharpshooters, and four companies of engineers.

The first two divisions will be deployed in two lines, with their artillery on the right, as shown in figure 23; the second line outflanking the first by the length of two battalions on the right of the first division, and on the left of the second. The two battalions on the right and on the left of the second line will be deployed, the others remaining in column as usual. The intervals between the ends of the two lines will be

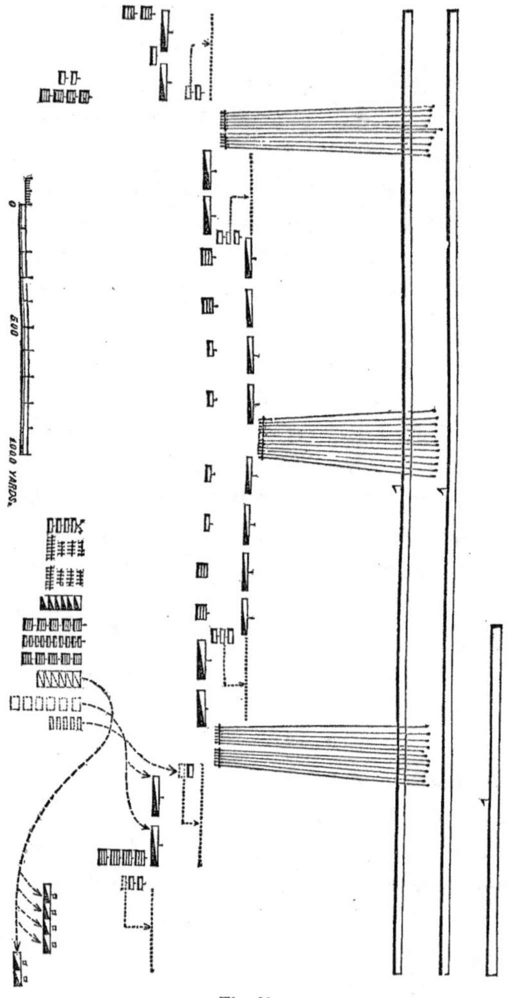

Fig. 23.

occupied by three companies of sharpshooters, in column, at the extremities of the first line; the two other companies of sharpshooters, in each division, will form a part of the second line, in the vacant space in the centre, occasioned by the second line outflanking the first. These companies, however, will only take up these positions after they have opened the engagement in the capacity of skirmishers.

The third division forms in echelon on the left, farther in rear than the second line, but will deploy only two battalions of its first brigade; the remaining two battalions of this brigade will remain in column behind the left wing, and the whole of the second brigade will be massed ready for movement. The artillery of this division will be placed in battery, on the prolongation of the second line, and two companies of sharpshooters will be placed as a link between the battery and the first battalion.

The reserve will approach the right, which seems to be threatened; the supplementary brigade of infantry will support the right wing, adopting the same formation as the brigade of the left wing. Of the five companies of sharpshooters accompanying this brigade, two will take position on the left, in advance of the brigade, and near the adjacent battery; the other three companies will remain with the column, and may be extended to the right as skirmishers. Half of the cavalry will form in echelon on the extreme right, which will complete the defensive arrangements.

Such dispositions as have been indicated, will only last until the moment when the enemy displays his designs, and will then be modified to suit circumstances. An attack upon the right is supposed, but the left is not left without resources. The latter is sufficiently strong to resist a serious attack until the reserve can be brought up; the second brigade of the third division is disposable for a flank attack upon any bodies of the enemy attempting to outflank the army on this side, or it may form squares, if circumstances require.

If, on the other hand, the real attack is made on the right, every thing is in readiness to receive it; the enemy finds four echelons before him, which compels him to extend his line very much, unless he prefers to attack one of the echelons, and then he is himself exposed to attack. If it becomes necessary to call the reserve into action, the disposable brigade of the left wing may replace it.

Although this order of battle makes sufficient provision for the security of the flanks, still it occupies a third more space than if the first three divisions were simply deployed in two lines, in the usual way. The army may thus resist four divisions of the same strength; and in order that the enemy may attempt to envelope it, with any chance of success, he must be in decidedly superior numbers. In the places where there is but one line of infantry, and a consequent appearance of weakness, the infantry is covered by a good line of sharpshooters, who restore the equilibrium. It may be observed that the defensive disposition, adopted by the army while in a state of expectation, presents a convex line to the front in its general outline. This form, to which the reasoning has led us, is the one which offers the greatest facilities for moving the reserves to threatened points, by following chords or radii, which are the shortest lines; the army on the defensive has thus the advantage of greater mobility. This fundamental principle is discovered in fortifications: every work, whose extremities do not rest on natural obstacles, should have an outline convex towards the front. But if this form is advantageous for providing against the preparatory movements made by the enemy for carrying out his plans, it is so no longer when the lines are actually engaged and fighting has taken the place of manœuvring; then the concave form is the best, because it is naturally enveloping, and delivers a converging fire. The skill of a general is shown in passing from one to the other form during the battle, by taking advantage of successes gained by one of the wings. When the ground presents firm points of support, it becomes possible to adopt the right line or the concave line in defence, as there is then no danger of being enveloped. The ground will always have a great influence upon the arrangements adopted, as well in the offensive as the defensive.

It is very important, whenever the field of battle is

obstructed and cut up, to make wide openings through hedges, &c., and to provide ample communications across ravines and small streams lying between different corps of the army. The existing roads should be repaired and widened.

Squares.—In the case now under consideration, squares are formed by divisions or brigades. Battalion squares are too small to serve as a basis for an order of battle, but they are excellent when a part of the line is strongly pressed. These small squares are rapidly formed, especially when the battalions are already in close columns, as usually happens; it is easy to place them in echelon or quincunx order, to flank each other. These battalion squares may be considered as a good manœuvre in certain cases; but when squares are the basis of the order of battle against an enemy whose principal strength is in cavalry, they should be formed of at least four battalions, that they may double their ranks without diminishing too much the interior space. This space is necessary for the reception of the staffs, and sometimes to contain even the cavalry when it is too weak to act.

This doubling ranks, when forming squares, is particularly applicable to those troops whose usual formation is in two ranks. It should not be done, however, by doubling companies or platoons, as there would be an improper mixture of subdivisions, and confusion would result at critical moments, a thing to be specially avoided. The doubling should be effected by placing the left half of each battalion behind the right half, or by placing even divisions behind the odd.

When the troops are thus formed in a doubled square, the front ranks of one or more faces may be detached without breaking the square. The angles *are the weak points*, and it is well to put the sharpshooters there in small solid squares, flanking the larger. The artillery is also placed at the angles, the pieces outside and the caissons inside; or, it may be in the middle of a face, or on the diagonal line, joining two adjacent squares. Several companies, distributed as reserves in the interior, will be very useful in re-enforcing points that are attacked, and giving support wherever necessary.

The cavalry is placed between the squares, so as to be flanked by them. From these positions it may fall upon the cavalry of the enemy, when disordered by the fire of the infantry; but when too weak to act, it takes refuge inside the squares.

Defensive Properties of Ground.—When the ground is not precisely what is called *a position*, but still presents some favorable features, they must be taken advantage of, no matter how insignificant they may appear, for whatever is not for us, in such cases, is against us. Thus, in 1690, Waldec lost the first battle of Fleurus, because he did not take possession of a plateau which had a very slight command of the field of battle. Marshal Luxembourg saw the fault, and at once took advantage of it, and, in consequence, gained a victory.

A simple wood, in which a few companies of skirmishers may be thrown, permits the line of battle to be extended, and a strong front to be presented, equal in extent to the enemy's. A piece of high ground gives advantages of another kind: if isolated, and of limited extent, forming a little eminence in the midst of a plain, it is proper, if there is no other reason to the contrary, to place the centre of the line upon it, as the enemy will generally be obliged to attack there, which will usually be unfavorable for him.

When, for any cause whatever, as, for example, to cover the line of retreat, it becomes necessary to have the high ground on the flank instead of in the centre, it should be held by one of the wings as strongly as possible. As it is advantageous to you, if held during a battle, so will it be of importance to the enemy to seize it. If the rising ground is of varying elevation, the highest is the most important point. Much will be gained if, in such a case, a block-house or redoubt can be erected.

The heights may form a line of hills whose direction is either parallel to the line of retreat or cuts it. What has been said above is applicable to the first case, and in the second the heights present in a greater or less degree the advantages of a good position. The first line will then be placed at the crest of the slopes; the second line and the reserves on the plateau or the reverse slopes, so as to be greatly or entirely out of view of the enemy. The line of battle thus follows the general direction of the chain of hills, and will be more advantageous, as it cuts the line of retreat more nearly at right angles. On the other hand, the more oblique the line of retreat, the less favorable is the position. The army A B (fig-

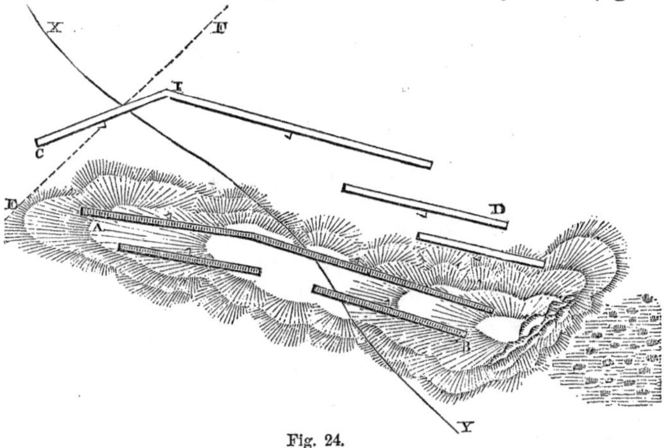

Fig. 24.

ure 24), occupying the heights whose general direction forms with its line of retreat, X Y, an obtuse angle, has its left flank very much exposed; for the enemy may envelop this extremity with superior forces by taking a position E F perpendicular to its line of operations; that is to say, by a natural deployment, and without an eccentric movement, it is ready to make a flank attack, which can only be counteracted successfully by changing the position of A B. This can only be made when the high ground presents a plateau of sufficient extent to permit a change of front, which is contrary to the supposition. In such a case it becomes necessary either to receive the flank attack or to abandon the high ground. It appears, therefore, that when there is high ground upon the line of retreat, it will not necessarily be proper for the army to occupy it and await the attack of the enemy, but the ground should have a suitable position and direction with reference to the line of retreat. The two lines should form nearly a right angle.

The danger just pointed out is not the only one. The enemy may take the position C I D, preferring the strategical to the tactical consideration. On his right, C I will suffice to guard his own line of retreat, while the main effort is made on the left. If this attack succeeds, it is evident the army A B is in great danger of being cut off from its line of retreat.

Use of Fortifications.—It is sometimes necessary to take positions where there is no natural support for the wings. Artificial ones must be created under such circumstances, if time permits. Very slight works of this kind, if judiciously constructed, may have a great value.

The engineer has sometimes also the important duty of fortifying a position selected and prepared for one of those mighty battles which decide the fate of empires. Time, that precious element in war, is then allowed him; he may then be extremely useful, if he has that eye for ground which enables him to make the best dispositions of his works, and that activity and devotion which ensures their execution. He should give ample opportunity for offensive movements, as a mere passive defence of intrenchments is injudicious. The works, whatever may be their character, should be separated from each other, leaving wide openings through which troops may debouch with considerable front. Each work should be well arranged, fraised, palisaded, and closed at the gorge, in order to offer a strong resistance. It is better to have a few works with a considerable relief and large ditches, than a number of insignificant affairs which would not check good troops for five minutes. Under protection of these works, and in a favorable position, an active defence may be made, which is so well suited to men of spirit, and tends so much to increase the *morale* of soldiers. This method is the best, because the impetuosity of attack is necessary for victory. A man feels inspirited when he is moving forward. To adopt a continuous system of intrenchments is to confess our own weakness, to render ourselves incapable of getting at the enemy, and to chill the ardor of the troops. The influence of fortifications in deciding battles is too well known to make it necessary to cite any examples of the fact.

Defensive Manœuvres.—We have hitherto supposed an army, obliged to act on the defensive, to have chosen a good position. with flanks well supported by obstacles, or by troops suitably posted. It may happen to be attacked in flank before its arrangements are completed. What shall be done then? It may either change front, pivoting on the least exposed wing, or a crotchet to the rear may be formed, or the second line disposed in echelon to outflank the first at the menaced end, or the reserve thrown upon the flank of the enemy making the attack. The last is the best course, as there is a boldness about it which is inspiring. It does not derange the positions of the several corps, that may then act to suit circumstances, under the protection of the attack of the reserve, which will surprise the enemy and check his movement.

The other plans are dangerous. The crotchet has the disadvantage of being easily enveloped, and having two long branches exposed to enfilade fire from artillery, which may prove very destructive. The faces are in such relative positions to each other that the troops cannot move forward without separating them at the angle and making an opening through which the enemy may advance, nor can they fall back without crowding. If the crotchet is formed under the fire of the enemy, confusion is the almost certain result. For a still stronger reason should a change of front of the entire army be avoided under such circumstances.

To move the second line by the flank towards the point attacked is scarcely better. The troops of the first line, seeing themselves unsupported, lose their confidence and do not hold their ground well. The two lines should be kept together as well as possible. Moreover, the second line may be outflanked as well as the first, and the movement referred to becomes impossible. It is only by arrangements made previous to the battle that the second line can be made to outflank the first by echelon. But at present we are speaking of movements made at the instant of fighting. It is prudent, then, to make no change in the primitive order of battle, and to leave the lines supporting each other.

The reserves alone are disposable for use everywhere. The reserves must now be thrown against the wing of the enemy. In the mean time, what are the other troops to do? They should effect a change of front, not in a methodical manner, as upon a drill-ground, where each corps must preserve its distance and alignments with the utmost rigor, but each will move by the shortest path to the point where it may soonest enter effectively into action. To explain by an example: take the line M, figure 25, surprised on its left flank by the army N, which prepares to attack. As soon as the enemy is discovered, the line M, which is supposed to be composed of four brigades, is broken to the left by divisions and formed in close column in each brigade. While these preparatory movements are going on, the reserve R moves to the position S, where it deploys so as to threaten the right flank of

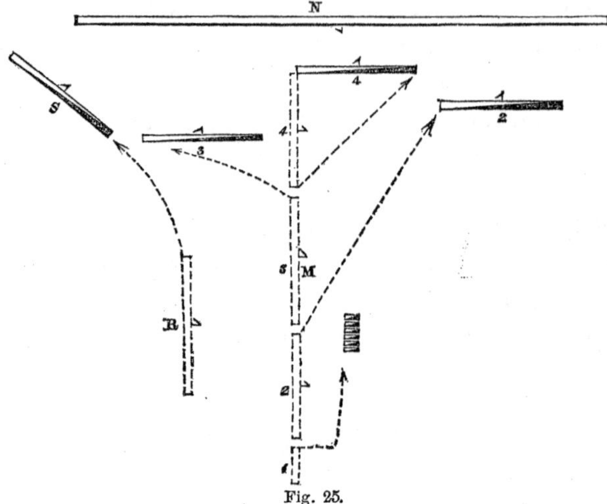

Fig. 25.

the enemy. The fourth brigade will be deployed upon its left division and the following brigade upon its right division in such a way that when the deployment is ended, which will require but a short time, these two brigades will be in echelon in front of the line of the enemy, and having a connection with the reserve. The second brigade moves off diagonally and takes position in echelon to the right of the fourth. The first brigade remains in column behind the centre, to act as a reserve. In this way there is very rapidly brought into line in the new direction a force equal to that in the former line, if we suppose the reserve to have been one-fifth of the whole.

If the army; N, extended towards its right still

more than shown in the figure, the reserve should not attempt to take N in flank, as this would separate it too far from the third brigade. It would then be proper for the reserve simply to deploy in front of N, but the second brigade would threaten the left flank of the enemy, and the general result would be the same in character.

For the sake of simplicity, we have supposed but one line. If there were two, the second would follow the movements of the first. The manœuvres we have just indicated are equally applicable to the case of a column suddenly arrested, while on the march, by an enemy in position. They are equally suitable for weak corps and for large armies. Thus, a battalion would first deploy two companies to receive the first shock of the enemy, while the remaining companies would be manœuvred into position. An inversion may be necessary, and hence the propriety of practising inversions at drill.

Retreat.—Whatever precautions may have been taken, or whatever courage displayed, it often becomes necessary to yield to mere numbers, or to the caprices of fortune. Happy, then, is the general who succeeds, after a battle long and severely contested, in withdrawing his army from the field in comparatively good order.

The movement to the rear is begun unnoticed, in a measure, the troops insensibly falling back as the enemy gains ground, on account of his superiority, which becomes more and more evident. Then comes the moment when longer resistance appears useless, and the general gives orders for the retreat.

Taking for granted that the army is drawn up in two lines, the retreat is begun by the first, which retires checkerwise; that is to say, the even-numbered battalions fall back to the rear some sixty or a hundred paces, while the odd-numbered battalions hold their ground. When the former have halted and taken position, and are ready to receive the enemy, the others retire in the same way, and so on.

In the mean time, the reserve, infantry and artillery, or part of the reserve, goes to the rear, to occupy the defiles the army has to pass in retreat. The second line regulates its movements by those of the first, sometimes hastening them, in order to gain some advantageous position where it may check the enemy temporarily. If the first line has suffered greatly, the second should replace it, a passage of lines being executed. It is better for the second line to pass offensively through the intervals of the first, than for the first to fall back through those of the second. Each battalion, being formed in a close column, is moved a few paces to the front of the second line, and rapidly deployed, or, better still, charges with the bayonet. The enemy is thus, in a measure, thrown upon the defensive, and becomes much more circumspect in his onward movement. The second method is dangerous, because the first line, coming back in a state of some confusion upon the second, may produce disorder in the ranks of the latter. The moral effect upon the troops of both lines is also bad.

In general, troops who are falling back should do so slowly and calmly, keeping their ranks well closed and in good order. They should frequently halt and turn to deliver their fire upon the enemy if he presses too closely. They may thus succeed, without very great disorder, in reaching some advantageous position, or being covered by the shades of night. Cavalry alone may retire rapidly; it is even its duty to do so, because in its rapidity of movement consists its safety. It should, however, return as rapidly to the charge, when rallied, in order to disengage the infantry.

When a regiment or a battalion is entirely broken, and the men absolutely running, the commander should seize a flag and plant it in some conspicuous place, at the same time causing the drummers to beat. The men, ashamed to abandon the flag they have vowed to defend with life even, will gradualy rally, and if the officers of all grades are active in reforming the ranks, order may be restored and confidence regained. Upon such occasions, the value of good officers is displayed, as well as the excellent effect of the example of old soldiers.

While the two lines are retiring the reserve does not remain inactive, but the infantry will occupy points where a firm stand may be made, or will form into squares, between which the army may fall back. Thus, at Marengo, the Consular Guard formed that famous square against which all the charges of the Austrian cavalry failed, and which gained from the army the name of the *granite square.* At the same time, the cavalry should make every effort to protect the retreat, by daring and repeated charges. It should never stop to count the numbers of the enemy, but should rush upon him, wherever he may be and however strong, in order to oblige him to deploy, and thus delay his movement forward. It is absolutely essential to safety to gain time, and nothing is more likely to effect this than dashing, furious attacks of cavalry, no matter what may be the result to itself. The artillery should cross its fire upon all points to be passed over by the enemy in pursuit, selecting favorable positions for this purpose. It must run some risks, even to the loss of a few pieces, as without its effective support the retreat of the infantry under the close murderous fire of the enemy must become a rout.

In the mean time, the baggage train should move off rapidly, under the escort of some troops. It has been kept at a considerable distance during the battle, and is set in motion as soon as the order is given to retreat. Its movement should be hastened as much as possible, in order to clear the way for the army.

Night at last comes on. The enemy, tired of fighting and exhausted by his losses, is forced to halt. This is the time for rallying and reorganizing the retreating army, and giving the men some food; but sleep is out of the question, as this opportunity must be used for getting the start of the enemy. A strong rear-guard, commanded by a bold and experienced officer, remains in position to cover the retreat. This rear-guard, when forced to retire, will do so slowly, and forcing the enemy to be cautious, by disputing the ground foot by foot.

The worst consequence of a defeat is not the number of men killed, and cannon captured, but the demoralization of the troops. Every means should, therefore, be used to remove their sad depression of spirits. The officers should wear cheerful countenances and speak encouraging words. The commander-in-chief, far from appearing disheartened, should seem entirely calm and collected; should visit the different bivouacs, and give his orders as if no reverse had occurred. He will thus impart to the men confidence in his firmness and courage. The troops will recover their spirits, and be ready to meet the enemy again. The victor of the preceding day may find, to his cost, that he did not know how to use his success, if he has permitted his army to sleep and his adversary to recover from his defeat. He will verify again the adage, that "in war nothing is done so long as there remains any thing to do."

Frederick the Great recommended that a beaten army should not retreat far, but halt at the first favorable position, to restore spirit and order to the ranks.

The Dukes of Weimar and Rohan, the two greatest generals of their day, after losing the battle of Rhinfeld, halted about fifteen miles from the field, and there rallied the remains of their army. By a night march, they presented themselves suddenly before the camp of the Bavarians, who did not expect such an attack, and were keeping a very poor lookout. They were surprised and routed completely, losing all their cannon and baggage. Here the Bavarians were accessory to their own ruin by sleeping on the field of battle, and giving themselves up to idle joy, instead of vigorously pursuing their beaten enemy. The rule laid down by Frederick, which he put in practice after the battle of Hohenkirch, accords with what has just been stated; but it is not always practicable for a defeated general to halt his army when and where he pleases. The more faithfully the army has performed its duty on the battle-field, the more difficult it is for him to do so; for the greater the obstinacy it has displayed, and the more determined the resistance, the less the probability of its withdrawing in order. It then becomes necessary to retreat to a considerable distance, to collect its scattered fragments and to receive re-enforcements.

We often see an army with one wing defeated, while the other still holds its ground and serves as a rallying point for the scattered battalions of the former, the whole then moving impetuously upon the enemy. Such conduct is often followed by complete success.

If, instead of that, the general orders a retreat, he may succeed in retiring without serious losses, but he is not the less certainly beaten. Indeed, victory is a prize sufficiently great to be renounced only after strenuous efforts to secure it. The first of duties is to fight well, and then to do whatever else is possible. A victory which is the result of a battle faithfully fought, may be equivalent to the destruction of the opposing army and the conclusion of the war. On the contrary, even in defeat, under such circumstances, honor is safe, and the vigor displayed in the fight and the losses incurred may dissuade even a successful enemy from engaging in similar contests.

CHAPTER V.

DEFENCE OF RIVERS AND MOUNTAINS—COVERING A SIEGE.

Art. I.—Defence of Rivers.

Disposition of the Troops.—In order to defend a river, the army should be divided into several corps of observation, placed in front of the points most threatened by the enemy, and not so near the banks as to be exposed to artillery fire. Each corps remains concentrated, only sending out small detachments to observe what is going on upon the opposite bank. Communication between the corps is constantly kept up by patrols. The detachments should be as few as can perform the duty thoroughly. They should be posted in clumps of trees and behind rising ground; in fact, wherever they can best observe the opposite bank without being seen themselves. The distance between the corps of observation will depend upon the breadth of the river and the time required for constructing bridges across it. The corps should be so near together, that either may receive the support of those next it in time to prevent the enemy from passing the river in its vicinity. If two or three hours are required for building the bridges, the corps may be from six to ten miles apart. Three corps of observation would thus watch twenty or twenty-five miles of the river, as the enemy could not attempt a passage above or below all three, or between any two of the corps, without resistance, first from one, and soon from two or three.

A strong reserve or principal corps should be kept to the rear, at a distance depending upon the front occupied by the corps of observation, and such that it can arrive at the point where the attempt is made to cross, before the enemy can carry out his object. If the part of the river to be watched is very long, two reserves should be formed instead of one.

To enable the corps to move readily from point to point, there should be a good road parallel to the river, and cross-roads connecting it with the position of the reserve. When the parallel road is very near the river, and under fire from the opposite bank, it loses its advantages, and a new one must be cut, farther to the rear.

Besides the general dispositions already mentioned, there are many details to be attended to. There should be a good system of signals, both for night and day; boats should be brought to the bank which is held by the army, and those that cannot be brought over should be sunk; the largest of them should be carried up stream, to be filled with stones and floated down to break the bridges of the enemy. If the river is fordable in some places, pits and trenches should be dug in the bottom, if possible; if not, the end of the ford should be obstructed by an abatis, or a field fortification of suitable character. If between two fords or two points favorable for passing the river, a narrow defile exists, it should be fortified and held, in order to control the road, and prevent the bodies of the enemy which may have passed, from effecting a junction without being attacked. When the bridges are of such a character as to be defended with difficulty, they should be cut, notwithstanding the opposition of the inhabitants of the country. In such cases, private interests must yield to the general welfare. As far as possible, however, such demolitions should be avoided. If a partial destruction will stop the enemy's progress, more than that is unjustifiable.

Bridge-heads.—If the bridge is to be preserved, it should be covered by strong field fortifications, while all proper precautions are taken upon the bridge itself, and behind it. A bridge thus held gives to the possessors the power of passing the river, whenever it is desirable. The fortifications erected to control a bridge constitute what is called a *bridge-head*, and should be arranged to suit the circumstances of the case, in accordance with the principles laid down in treatises on the subject of Field Fortifications.

The river, instead of crossing the line of operations of the enemy, may be parallel to it. In this case, a bridge that is to be held must be defended by fortifications on both sides, as it is impossible to say on which side the attack may be made. There is thus formed a double bridge-head. Such a work gives great advantages to the party on the defensive, if it manœuvres rapidly, for there is always a safe passage from one bank to the other, and the bridge-head may be occupied by only a sufficient garrison to hold it against the enemy; this garrison may be relieved or re-enforced at pleasure, so as to make a very efficient defence. The enemy cannot safely pass by, leaving such a point upon his line of operations. He must therefore attack the fortifications, and may thus be forced to a considerable loss of time, men, and material. If he divides his forces to surround the work, and attack both sides at once, the army on the defensive may concentrate in force against a fraction of the enemy and destroy it. In order to operate successfully, therefore, against such a work, the enemy must have a very great superiority of force.

The Archduke Charles remarks: "There is no better defensive position than one which keeps the enemy constantly apprehensive of being attacked. An army behind a river, with strong bridge-heads, is in precisely such a position. Almost any other may be turned, but a bridge-head, well built, secure from capture by assault, and perfectly covering a passage of the river, need never suffer from an insufficiency of numbers in the garrison, or from want of provisions and munitions of war. The enemy is forced to watch carefully such a communication, by which superior forces may be at any moment thrown upon him."

When a simply passive defence of a river can alone be made, which is an unfortunate state of affairs, special attention must be given to those points that offer the enemy advantages for building bridges. Such are those where the river makes an elbow convex towards the enemy, as batteries may be constructed to cross their fire in front of the ground where his leading troops would be landed. Other favorable points for him are those where the river is divided into several arms by islands, that may conceal the preparations made for passing; also the mouths of tributary streams, by means of which the enemy may bring up from the interior boats and other materials for bridges. Batteries should be constructed at such places, so as to sweep with their fire the opposite banks, and the ground adjacent. When there is no time to finish in a complete manner such works of fortification as may be suitable, simple trenches and rifle-pits should be rapidly built, so as to envelope the space upon which the enemy must place the foremost of his troops, and expose them to a close converging fire. Upon one occasion Eugene had gained a march upon Vendôme, and was attempting to throw a bridge across the Adda at a very favorable spot. Vendôme came up as soon as he could, and arrived before the bridge was completed. He tried to arrest the work of the pontoniers, but in vain. The ground was so well swept by the artillery of Eugene, that he could not get near enough to injure the workmen. Still, the passage of the river must be prevented. Vendôme put his army to work upon a trench and parapet, surrounding the ground which the Imperialists must occupy after crossing. They were finished nearly as soon as the bridges; Eugene deemed the passage of the river impracticable, and ordered a retreat.

When the attacking party is thus caught in the act of constructing his bridges, he is taken at great disadvantage, because the main body of his army is not across, while the few troops who have passed in boats are not in sufficient numbers to force their way forward, and make room for the army to debouch from the bridges and deploy in line of battle, even if it be practicable for it to do so under the converging fire of the artillery and infantry of the defenders.

The enemy is also in a perilous condition, if a portion of his forces have crossed, and the defenders succeed in breaking the bridges, and thus cutting his army in two. Every effort should therefore be made to do this. If large boats, filled with stones, and heavy rafts, constructed of trunks of trees, be sent down the stream, there is great probability, if the current is rapid, that some, or all of them, will strike the bridges and break them. In 1809, the Austrians succeeded in cutting the French army in two, by throwing into the stream several wooden mills which were upon the banks of the Danube. The corps which had passed the river, and taken possession of Gros-Aspern and Essling on the left bank, were surrounded by the whole army of the Archduke Charles, and obliged, after a heroic but ineffectual resistance, to retire to the large island of Lobau, that here divides the Danube into two channels. After this check, two months were consumed by the French in immense and persevering labor, in building several strong bridges, and preventing them from rupture by stockades above them.

Secondary Means.—It is very important for an army engaged in defending a river not to be deceived by false demonstrations, and to have timely warning of a real attempt at a passage. Officers of experience should be sent to the posts of observation, who are not easily deceived by feints. Signals must be agreed upon, by which timely notice may be given of suspicious movements of the enemy.

There is, moreover, a military law, a law which can never be violated without dishonor, that requires a commander to go in the direction of a cannonade when it is so near and so prolonged that there can be no reasonable doubt in his mind of a serious engagement being in progress. No excuse can justify him for remaining unmoved in his position when the thunder of the cannon tells him that a battle is going on at no great distance. He cannot even plead the orders of a senior, for contrary orders may have been sent but intercepted, or their bearers may have been killed.

He must decide as to the urgency of the case, and take the responsibility of moving where honor and danger call. If the soldier and subordinate officer should render implicit obedience to the orders of their superiors, the case is different with a chief of long experience in war; he must necessarily exercise a certain degree of discretion in the performance of his duty, especially when an unforeseen case arises. Unity of effort towards a common end is a fundamental principle in the operations of war, and this requires movement towards the sound of the cannon, when there is doubt in the mind of the commander.

Movements of Troops.—As soon as the alarm is given, the posts of observation nearest the point of passage hasten up and charge the troops already over, whatever may be their numbers, to drive them into the river, if possible, before others can cross to their assistance, or at least hold them in check until re-enforcements arrive from the rear. If the enemy commits the fault of commencing the construction of his bridges before he has possession of both banks, the sharpshooters may draw near and pick off the pontoniers with ease.

The artillery and cavalry ought to be the first to arrive to the support of the advanced posts. If the ground favors, the cavalry charges vigorously, while the artillery takes position to reply to that of the enemy, or to crush the troops already over. The horse artillery may be very useful on such occasions, from the rapidity with which it may be brought upon the ground and placed in position. The corps of observation, on the right and left, are put in motion immediately, and lose no time in coming up and taking part in the action, which becomes continually warmer. The position of the enemy becomes more critical, as he can scarcely use his artillery for fear of firing upon his own troops. He will be unable to hold his ground, pressed and surrounded as he is, if he has not succeeded in completing his bridges by the time the reserve comes up. The troops that have crossed must lay down their arms.

It often happens that the defenders are taken by surprise, notwithstanding all their precautions, because the enemy may deceive them in a thousand ways. The troops will then reach the ground too late, or, coming up in successive portions, may be destroyed. The passage is forced, and, as the assailant is usually the stronger in such cases, the other army must fall back to some other position in rear.

Whatever may be the danger of the passage of the river by the enemy, there is no reason for occupying a great extent of the river to prevent it, and thus spreading out the troops in a long weak line. The enemy must do one of two things. He must either keep his forces united, or divide them. If he concentrates, the army on the defensive should pursue the same course, remaining opposite to him and gaining all possible information of his movements. It will be difficult for the enemy, in such a case, to effect a passage by surprise, unless the locality is very favorable to him, and information of his movements cannot be obtained, either through spies or the inhabitants of the country. If he occupies a great extent of country by separate corps, he need not be feared at any point. The army on the defensive should be kept together opposite the centre of his line, thrown upon any corps that might succeed in crossing, and crush it before the arrival of other detachments, which, in the case supposed, are quite distant. By concentration, is not meant crowding the whole army in a single camp. It would then be impossible to watch the banks of the stream. The troops are sufficiently concentrated when the corps are near enough to afford mutual support; and it is quite certain that when this is the case, the enemy cannot effect a passage between two of the corps before they can resist him with effect. The word *concentration* must, therefore, be not understood in an absolute sense, but receive a liberal interpretation, as, indeed, is the case with all military terms.

A very effectual method of disconcerting the enemy, and neutralizing the moral effect of the successful passage of a river by him, is to cross at some other point and begin offensive operations upon his territory. Such a determination upon the part of a general would show him to be a man of spirit. The imagination of the troops is excited, their hopes aroused, and their courage renewed, while, at the same time, the enemy is astonished and demoralized at the very natural supposition that the opposing army is in greater force and better supplied than was believed. The enemy will speedily repass the river, to protect his own territory. In 1674, Montecuculi crossed the Rhine. Turenne at once crossed in the other direction, instead of seeking to defend the frontiers of France by the usual means. The initiative that Montecuculi had so skilfully taken was of no avail, as he was obliged, by the still more skilful manœuvre of Turenne, to return to the right bank of the river.

Sometimes the defenders withdraw designedly from the river, in order to entice the enemy across, and then return upon him before his whole army is over and in position. The general who retires in this way is only justified in so doing by the character of the ground, which permits him, upon his return, to have a good position, while the enemy is crowded and unable to deploy his forces. Unless these are the circumstances of the case, he runs the risk of a defeat if the enemy succeeds in throwing over troops enough. Such a mistake has been actually made. Marshal Créqui, at Consarbruck, suffered the enemy to cross, deferring his attack with the expectation of enveloping and routing a greater number. To those who were surprised at his course, and expostulated with him, he replied, that the greater the number who passed, the more decisive would be their defeat.

However, so many passed that he could not resist them, and the marshal was completely beaten and covered with shame. This mistake made a fine general for France, as Créqui was cured of his rashness, but still retained ardor enough for great undertakings. We may learn, also, from the life of Créqui, that a man who is truly great, and worthy to command others, knows how to profit even by his own faults.

Art. II.—Defence of Mountains.

Such a country is best defended by rapid manœuvring and energetic attacks upon the enemy. His plans are thus thwarted, and he is obliged to think about his own safety. He has not the same freedom of movement as the defenders, because he is obliged to protect and keep open his line of operations, whilst they are at home and find a line of retreat in any direction. The inhabitants of the country are also generally ready to aid the defenders, and fall upon the assailants at all favorable opportunities. They will at least render important services by watching the movements of the enemy and giving notice of them. Some of them will always act as partisans, and do much valuable service.

The first rule to be observed in defending a mountainous region is not to try to close all the passes, as an attempt to do this would lead to injudicious scattering of the troops, and a very weak front is offered to the enemy at every point. The plan of concentration should be followed, as far as the country will permit. Those positions should be strongly held, from which troops may be rapidly moved to any point threatened by the enemy. It is possible to assemble several battalions only in the valleys; these, then, will usually be the places where the defence will be mainly made; but the advantages given by the higher ground in arresting the movements of the assailants must not be lost sight of.

In the next place, there are two things to be done in a warfare of this kind: *to manœuvre offensively*, in order to dislodge the enemy, take his columns in flank, separate his detached corps, seize his convoys, &c.; and *to fight defensively*, that is to say, to take position in the most advantageous way possible whenever it is necessary to engage seriously. Although these two precepts may seem contradictory, a skilful officer will know how to apply them together; he manœuvres to get upon the flank or rear of the enemy, and having succeeded in doing so, he takes a position, or, at any rate, does not attack unless he has the decided advantage over the enemy. The latter, being unable to move farther without danger of having his line of retreat cut, is obliged to turn upon the corps which is in position and attack it. Until that is done, he is in constant danger of being separated from his re-enforcements and means of subsistence. He must, therefore, be the cost what it may, march upon this position and attack it before advancing to other operations. To oblige him to do this was the object of the manœuvre; if his attack succeeds, he will at any rate suffer greatly; and if he fails, his position is critical.

Artificial Obstacles.—Besides manœuvres of the troops, whose importance is undoubted, there are other means, not to be neglected in a good defensive system; these art provides, for the purpose of delaying or completely arresting the movements of the enemy in certain localities, and for strengthening positions for engagements. Thus, preparations should be made in advance for obstructing roads. Mines will be placed in the piers of stone bridges, and combustible materials got in readiness for burning those of wood. Parts of the roads that are narrow and cannot be turned will be blown out with powder, and temporary bridges made, which can be readily destroyed. At other places mines will be prepared to throw down masses of rock in the roads, either in anticipation of the approach of the enemy or when he is passing, and thus be the occasion of much damage to him. Abatis may be formed across narrow roads, or barricades constructed of earth, timber, or rocks. These measures suppose some haste in their execution; but, when time is abundant, the better plan is to build block-houses or larger earthen works in those parts of the valleys which are commanding from their position, and yet so contracted as not to require works of great extent. A strong redoubt, well fraised and having a considerable command, is a difficult obstacle for an enemy to pass in such a locality as is here spoken of.

Positions.—Good positions, well defended, will always give the enemy much trouble in mountainous warfare. He can take them by assault only, at great loss to himself, and much time will be consumed if he attempts to turn them.

Positions will be found upon the high ground or in the valleys. The first, although ordinarily strong against attack in front, are often easily turned. It is, moreover, generally difficult to furnish food and water to the troops holding them, and they should, therefore, be held by only sufficient forces to repulse the enemy with loss if he attacks in front. Small bodies will retire with facility and without fear of being overtaken; they will take every advantage of the ground to check the enemy, and will give themselves up more to fighting than to getting out of the way. Finally, if a small body is unfortunately captured, the moral effect is not near so great as would be produced by the loss of a large detachment.

The detachments that defend the heights, usually belong to more numerous corps encamped in the lower ground. The troops should not be scattered around, holding every little by-path, as this will prevent a strong resistance being made at any point. The detachments will be grouped near the main pass, which they should hold long enough to give time to the troops below to come to their support. By placing reserves in rear and on the flanks, at points where the lateral roads come in, they will guard against being cut off. The natives may be very useful in giving warning by signals of the approach of the enemy, and a few of them, with arms in their hands, will guard the by-paths as well as the best soldiers, and with more confidence, on account of their being perfectly acquainted with the country, and certain of their ability to escape at the last moment.

The best positions are usually in the valleys, because they are generally better supported, being occupied in force, and with the high ground on the flanks guarded. The enemy cannot turn them so easily as those upon the hills; if he wishes to do so, he is obliged to make wide détours through other valleys. Positions, properly so called, are comparatively few. They are either across or along a valley.

When the defence is made while ascending a valley, the enemy attacking from below, the ground, without presenting what are called positions, is nevertheless favorable almost everywhere. But when the top of the ascent is passed, and the defenders are descending while the enemy is higher, he has the decided advantage. It is only at considerable intervals that such elevations in the valley occur as to form defensible positions; but these are generally very good, because contracted in extent, and having the flanks upon the precipitous sides of the mountains. These positions often have the disadvantage of being cut in two by the stream which flows at the foot of the valley, and, if it is not fordable, the wings cannot support each other. Thus the enemy may attack one wing with superior numbers, and, when it is driven back, threaten the other in flank and oblige it to retire also. Such positions, therefore, to be available to the defence, should offer free communications between the different corps of the army, while those of the enemy are difficult or impossible. With this view, the bridges above the position should be cut, while those upon it and below should be retained and even new ones constructed. Such is the position across a valley.

A position is taken up along a valley, when it is desired to arrest the progress of an enemy wishing to emerge from a cross valley. The ground will usually rise like an amphitheatre in front of the mouth of the transversal valley, so that the heads of the columns of the enemy, in debouching, will be exposed to a cross and enfilading fire, and it will be extremely difficult for him under these disadvantages to advance to the attack. But, in order that the position be thus favorable to the defence, and that the general may expect to force the assailant to retire, the valley must be so narrow as to enable troops on the opposite slope to reach with their fire the enemy emerging from the transversal valley. If the main valley is too wide for this plan to be pursued, the army on the defensive must be on the side next the secondary valley, and must act offensively, enveloping the enemy as he debouches. There will not be the same advantage of position as before, but the defenders are enabled to bring fresh masses of troops against those of the enemy, who are necessarily much strung out, and perhaps fatigued by painful marching. When an attack is thus made upon an enemy seeking to debouch from a secondary into the main valley, the artillery ought to be placed in front of the infantry; this is in violation of the ordinary rule, but it is permissible, because the enemy, in such a case, will usually have little or no artillery in action; and it is necessary, because the space required for the ordinary dispositions is wanting, and the artillery must be in front to attain the enemy at the greatest possible distance.

Intrenchments.—There are some examples of valleys defended by intrenched lines. These are, undoubtedly, good when the valley is narrow, and the mountains on the flanks very steep and only turned with great difficulty; when the intrenchments have masonry revetments, and the ground on which they are built has such a command over the valley that the artillery can sweep it to a great distance. These conditions are not so often satisfied as is supposed, and hence the proper use of this kind of defences is not frequent. Reference is not now intended to small redoubts for detached posts, nor to the larger earthworks often advantageously used in mountain warfare, but to those extensive lines intended to close up an entire valley; such lines, to have any real value, must fulfil the conditions laid down above.

Movements of Troops.—When all hope of a successful defence of a valley is lost, the commander may attempt an offensive movement through a lateral valley, rather than retire immediately into the interior. Nothing can be more honorable and brilliant than such an effort, and it is often crowned with the success Fortune loves to bestow upon daring actions. A timid chief, who thinks only of defending positions, and who falls back constantly as the enemy advances, deserves and receives none of her favors. In warfare in a mountainous country some risk must be run in making eccentric movements upon the flanks and rear of the enemy, because the character of the topography favors; the body of troops making the movement is not likely to be enveloped, even when very inferior in numbers, because the enemy, usually spread out in the valleys over a great length, can with difficulty assemble his troops, and, even when that is practicable, he requires some time to reach the high ground and rid himself of the restraint imposed upon his actions by want of space. A small body, well posted, will perhaps fight all day before the enemy can reach its flanks, so difficult will he find it to make combined movements when attacked thus unexpectedly. The paucity of communications is unfavorable to the prompt transmission of his orders, and when, at last, they reach their destination, the time for their execution may have passed. The attacking party has, therefore, many chances of gaining some signal success, and the fame of it, increasing as it goes, will make the enemy uneasy and encourage the defenders. The worst thing that can happen in such a case is to be obliged to fall back after an unsuccessful attack; but retreat is easy and not very dangerous, as the troops that retire are getting constantly nearer their supports, while the enemy is leaving his behind. Here is another example of the modifying influence of localities upon general rules.

When a defensive corps has succeeded in regaining some important pass or high ground, by which means the communications of the enemy are interrupted, it may either take the offensive or remain in position. The first course will be pursued if the enemy is not in great force and the ground is favorable; in the other case, the defenders will simply hold on to what has been gained, and make preparations to receive an attack, by occupying woods and commanding ground with sharpshooters, by making abatis across the roads, and collecting large rocks and trunks of trees to roll down upon the assailants. The enemy will make a powerful effort to reopen his line of communications; if it is unsuccessful, the assailed should not give themselves up to an imprudent pursuit, but should reoccupy their position, and prepare for the enemy a similar reception whenever he advances.

A great advantage is gained in an attack in such country by occupying a height which had been deemed inaccessible, and planting a piece or two of artillery upon it, even if of small calibre. The enemy is disconcerted and alarmed, and inclined to abandon his position at once, especially if attacked in front. Thus, at the combat of Val Carlos, General Noguès had a piece of artillery carried by main strength up very steep slopes to the heights which commanded the camp of the Spaniards. The latter were astonished, as they had supposed the heights inaccessible, and they rapidly retired from their position before an enemy greatly inferior in forces.

It has been already seen how the character of the ground will modify general rules. Still another example in point may be mentioned. The combined movement of several detachments to envelop the enemy in an open country is very properly condemned; but, among mountains, such a manœuvre may be safe and very successful. The enemy is unable to interpose between the detachments and beat them separately; each should be strong enough to defend the valley it occupies, and it is then in no danger of being surrounded; moreover, all the detachments can retire safely if they fail in their attempt. They should have, at least, mountain artillery with them, not only for the effect to be produced upon the enemy, but to make signals. This is often the sole means of communicating between the detachments.

If a detachment of the enemy is met in one of the valleys, it should be attacked in front, while infantry, in dispersed order, attempt to gain the high ground on the flanks and rear. This precaution should be taken in all movements along valleys.

With the precautions that have been mentioned, boldness may be displayed, even to rashness, and will generally lead to success. The Archduke Charles says: "Nowhere are such wonders accomplished by boldness as in broken countries, and especially among high mountains; where the warfare is necessarily one of posts, the engagements are generally unexpected, and the enemy, surprised by unusual daring, is paralyzed at the critical moment." In combined movements, rivers and lakes may furnish a very excellent means of communication. The transports should be propelled by steam, and armed, or escorted by gunboats.

If, after disputing the upper valleys foot by foot, by taking successive positions in them, or by manœuvring to separate the columns of the enemy and divert them from their object, it becomes necessary to abandon them, the lower valleys still remain to be defended, where larger bodies of troops can be concentrated, and the ground offers many positions where a good general may use his tactical ability most effectively. A point will be selected, for example, at the junction of several valleys, or at the mouth of a single one. The defensive army will have great facilities for moving rapidly from one outlet to another by the shortest lines, while the enemy can only operate in a single valley, or cannot communicate between adjacent valleys except by great labor or wide détours. The plan then will be to retard him simply in the narrow valleys, which can be done by small detachments, while the main body attacks where a greater display of force is practicable. After beating the enemy here, a rapid movement may be made to the outlet of the nearest valley, where it is more than probable a similar success may be gained.

Suppose, for example, that an army of 30,000 men is advancing by three valleys towards an objective point, M, figure 26, which is probably the principal city on the outskirts of the mountains. The roads through these valleys converge towards M, but the two on the right unite before reaching M. The invading army, in order to occupy these three valleys, is obliged to send an equal force of 10,000 men into each. Suppose the defensive army to contain about 19,000 men. They may or not have been also divided into three equal parts, but the moment has now arrived for a decisive concentration; 3,000 men can check 10,000 in each valley, or at least delay them sufficiently by cutting bridges, closing defiles, taking flank positions, &c. The general will, therefore, leave three detachments, *a*, *b*, *c*, each of 3,000 men, in these valleys, and with 10,000 men will take up a position *d*, near *m*, where two of the roads meet.

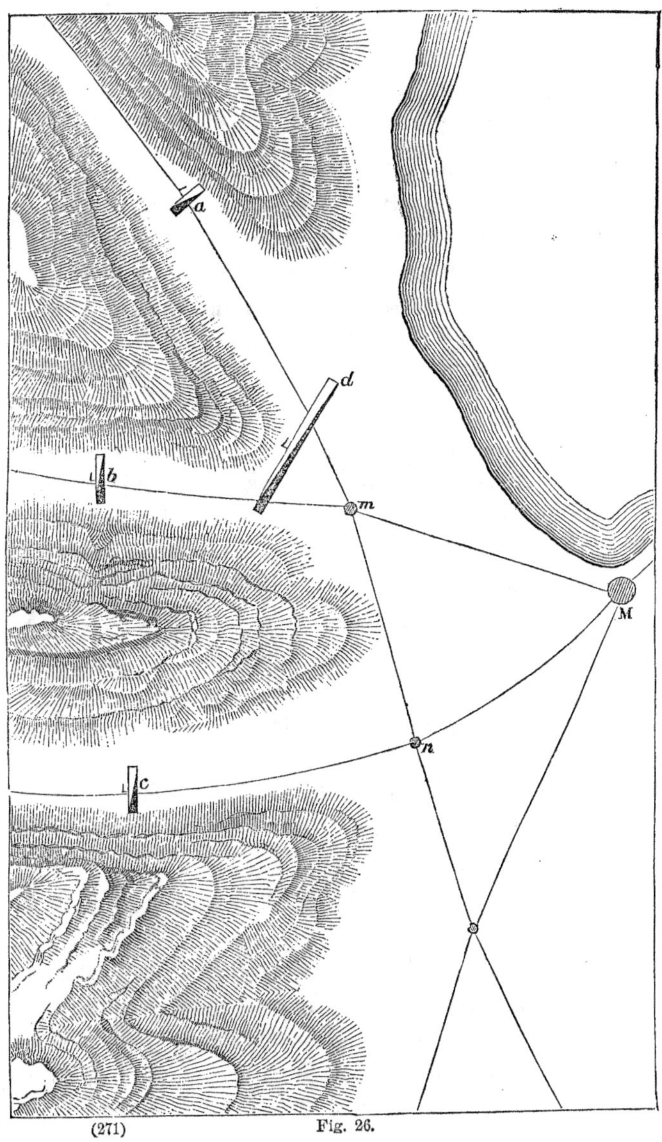

(271) Fig. 26.

If the valley on the right is most accessible, he will strike his first blow there. He will inform the commander of *b* of his intentions, ordering him to maintain his position firmly, while he joins *a* and fights a battle. He will have 13,000 against 10,000 of the enemy's forces, and the result cannot be very doubtful, especially if the ground favors his operations. If the enemy retires without a battle, the news must be proclaimed far and wide, but the general should not engage in a pursuit by which he would be removed too far from *m*, which is now the essential point; he will, on the contrary, after a show of pursuit, return to his former position, to manœuvre from it as circumstances may justify or require.

If the enemy fights and is beaten, the corps *a* should be sent in pursuit, while the main body hastens to *b*, to fight again with a similar superiority of forces. Immediately afterwards the main body *d* moves by the cross-road *m n* to the third detachment *c*, *b* having been sent in pursuit of the enemy in the middle valley, or ordered to hold him in check. Upon this last occasion the victorious army will follow up its advantage, pressing the third corps of the enemy vigorously, and gaining the flank of the other two, which will thus be forced to retreat.

If, on the contrary, the enemy does not advance along several valleys, as has been supposed in the example above, but by a single road, with a view of penetrating into the interior of the country, his forces will be concentrated, it is true, but he will want room to use them efficiently. The defenders should not be alarmed at this concentration, but bravely prepare to meet the enemy at the mouth of the valley. Such a position and order of battle will be selected that the enemy will be enveloped as he debouches, and exposed to powerful cross-fires. Courage and firmness are especially necessary now, for if the defensive army is beaten in this position, with the ground favoring it so much, it will not probably be successful elsewhere.

Communications.—The preceding remarks demonstrate the necessity of good roads for the defence of a mountainous country, as without them it would be impossible to move rapidly to threatened points, to effect the speedy concentration of troops when desirable, and to make those offensive returns which have so excellent an effect in raising the *morale* of the army and disconcerting the enemy. If possible, therefore, a system of roads should be prepared in advance for the defence of a mountainous country. At ten or twelve miles to the rear of the principal crests of the chain, a road should be built parallel to it, passing over the spurs and crossing the perpendicular roads, thus giving an opportunity to move freely from right to left in case of need. Each of the perpendicular roads should be closed by a strong fort, well located between the crest of the chain and the parallel road. Still farther to the rear, at the foot of the mountain, a second parallel road should be built, crossing the outlets of the valley. By means of such a network of cross and longitudinal roads, the defence might be made as active as possible. If such an arrangement is impossible, on account of topographical difficulties, the spirit of it, at least, may be attained.

Summary.—From what has been said on the subject of conducting a defensive war in a mountainous country, it appears it should be characterized by peculiar boldness and activity. Every thing depends upon rapidity of movement and the art of acting offensively, even if upon the defensive. The enemy cannot be driven out by attacking him in his chosen positions, but by manœuvres to turn his flank and get in his rear; by obliging him to fight offensively and upon unfavorable ground; in a word, the course to be pursued is, as before stated, *to manœuvre offensively and fight defensively.*

Art. III.—Covering a Siege.

While a portion of the army is besieging a fortified place, the remainder repels the enemy, if he attempts to succor it. The *besieging army* performs all the actual labor of the siege; the covering army is called sometimes an *army of observation.* The former is established in camps, near the besieged place, and beyond the range of its cannon; the camps are fortified both against sorties from within and attack from a succoring army. The details of siege operations are described in treatises especially devoted to such subjects. We will simply state some rules for the government of the covering army.

It should not be too distant from the besieging army, in order that, in case of need, re-enforcements may be drawn from the latter, who will return to their camps after the necessity for their assistance has passed away. Such re-enforcements, arriving at an opportune moment when an engagment is imminent or in progress, will be of great value, and will have a powerful effect in defeating or repulsing the enemy. When Bonaparte was besieging Mantua, he did not limit himself to drawing several battalions from the besieging army when about to engage the numerous forces that seemed to surround him, but he brought up the whole besieging army, and, uniting it to the army of observation, he gained the battles of Lonato and Castiglione. But it is disadvantageous to be obliged thus to withdraw the whole besieging army, because the siege-works must then be suspended and the siege artillery abandoned; the latter can be only recovered by defeating the succoring army and capturing the besieged place. If the covering army is too distant, the enemy may, by a rapid movement, come unexpectedly upon the besieging army, which is often, indeed generally, in no condition to fight a battle. The consequence of such an occurrence would be the raising of the siege and the abandonment of all the siege materials. In such a case the covering army should rapidly follow upon the heels of the enemy and attack him in rear, while the besieging army should concentrate and attempt to hold its ground by the aid of the fortifications already erected around the camps. There is no other way of recovering from the injury caused by the rapid march of the enemy.

The covering army will remain in a fortified position only when it is accessible in a single direction. If the enemy can pass along some other route, or can come up by several roads, they should be all watched by detachments, while the central mass, far from being tied down in intrenchments, should be as easily moved as possible; with this object the commander should send to the camps of the besieging army all unnecessary baggage and all the sick and wounded; his artillery should be in perfect marching order, the roads should be repaired, &c. But if the main body, in the case supposed, should not make use of intrenchments, the detachments of observation should by all means do so, when the ground favors; defiles should be fortified and bridge-heads constructed, to enable these detachments to resist superior forces, and for a time even to check the entire succoring army.

Scouting parties should be sent out to a distance, and frequent reconnoisances made, to ascertain what the enemy is doing or going to do. If he is concentrating his forces, an attack may be expected, and preparations must be made accordingly; if he is collecting wagons and provisions, he is about to attempt to throw supplies into the besieged place; the besieging army should be informed of this design, and arrangements made to capture the train, or at least prevent its passing in.

If several roads lead from the enemy's positions to the city under siege, the covering army occupies the central one, defending the others indirectly, by the fear occasioned the enemy of being taken in flank or rear, should he attempt to pass along one of the lateral roads. Here, as in every other case of defence, dispersion of the forces, that inevitably leads to partial reverses or total ruin, must be carefully guarded against.

When the enemy has developed his designs, the covering army should march to meet him and attack resolutely, whatever may be the relative forces; this is no time for counting numbers. In the mean time, the noise of the cannon, staff officers, and couriers will have given notice to the besieging army of what is passing. Its commander will call out his whole disposable force, after making sufficient provision for guarding the trenches, and taking proper precautions for repelling sorties from the place, which are more to be apprehended at this than at any other time. The troops intended to participate in the engagement should lose no time in leaving, and when they arrive upon the field of battle, they take their places in the line, or form a reserve, or fall upon the flank or rear of the enemy, according to circumstances. Nothing can have a more powerful effect in procuring a victory than the appearance of such re-enforcements upon the field of battle at a critical moment in the contest. Desaix's division, suddenly debouching upon the field of Marengo, regained the day for the French. If, notwithstanding the assistance rendered by the besieging to the covering army, the latter is obliged to fall back, it should endeavor to do so in good order, and take position at a little distance from the field of battle, in order to keep the enemy in a state of uneasiness, and to prevent him from falling with all his force upon the siege-works. The covering army will be put in as good condition as possible, calling in all detachments at some distance, and if it is then sufficiently strong, another attack may be made upon the succoring army, which will have been during this time arrested by the intrenchments of the besieging army. Then, if both parties are equally determined, will be witnessed the singular spectacle of two armies at once besieging and besieged.

If the covering army has been unable to resist the succoring army, and the latter has immediately attacked the lines of the besiegers and forced them, affairs are certainly in a bad state, and the siege must be raised; but all is not yet lost. The remains of the two armies must be rallied and a new one formed, which will soon be in condition to fight again.

If the enemy is fearful of compromising himself, and remains closely within the walls of the fortifications, no attack can be made upon him; but many mouths to be filled soon exhaust the supplies, and the place will capitulate for want of provisions. Every thing that had been previously lost will now be recovered, and at the same time the result is obtained without loss of life—a most pleasing thing to a chief who loves his men and is averse to shedding their blood.

If, on the contrary, the enemy has more confidence, and remains in the open field rather than shut himself up in the fortifications, new combats will follow, and the effort must be made, by partial successes, or by the gain of a battle when the enemy will fight one, to force him into such a position that he may be blockaded. If, finally, the enemy divide his forces, leaving a part in the place for its defence, and with the remainder takes the field, so much the better for the other party, as there is an opportunity offered for cutting off from the garrison the portion in the field. The attempt should be made to slip in between the two by a night march, such precautions having been taken that the corps in the field may be attacked, thoroughly beaten and dispersed, or captured, before assistance can arrive from the place. Thus, Marshal Soult, at the siege of Badajoz, having to deal with an army stronger than his own, skilfully took advantage of an opportunity that offered of restoring the equilibrium. Ten thousand Spaniards, to avoid being too much crowded, passed out of Badajoz and encamped upon some high ground, which was separated from the French army by the Guadiana and covered by the

Gebora. Fire was opened upon the Spanish camp from the French camp, with long-range howitzers, with a view of obliging it to be pitched as far as possible from the outworks of the place. An hour before dawn, a passage of the Guadiana was effected in boats, the Gebora was forded, and, while Marshal Mortier made an attack in front upon the high ground, and sent his cavalry to turn the right, two or three thousand infantry were posted in the valley between the fortress and the camp. The Spaniards were thus completely cut off. Eight thousand prisoners were taken, five or six hundred killed, and the rest escaped. This was the brilliant combat of Gebora, fought February 19, 1811.

CHAPTER VI.

COMBATS AND AFFAIRS.

The term *combat* is applied to a partial engagement—to a conflict of parts of two opposing armies. Generally, the losses in combats are proportionally greater than in battle; often, when the forces engaged are numerous, the action is called a battle, although more strictly a simple combat. But too much importance must not be ascribed to mere definitions, for the name has really no influence upon the thing; a combat is a small battle, and a battle is a great combat.

A combat may occur between two bodies of infantry, between two bodies of cavalry, or between a body of infantry and one of cavalry, between troops with artillery and troops without artillery. It may occur in the open field or in intrenchments, &c. It is well to examine these different cases separately, although it generally happens that they all may be discovered in a single engagement.

Art. I.—Combat of Infantry with Infantry.

A combat of infantry against infantry presents much the same appearance as a battle; skirmishers in front to open the engagement, the first line deployed to deliver its fire, the second line covered by accidents of the ground, and out of range of fire-arms, several masses upon the wings and in reserve, to guard against turning movements. If the troops are few, they will form a single line with a reserve. In such an engagement, the best armed men and the best marksmen have a great advantage, for the opposing lines are not very distant, and the firing is kept up a long time, as there is neither cavalry nor artillery to be feared.

When several battalions are engaged, some of the manœuvres prescribed in the works on tactics may be applied, and may lead to important results for the party that knows how to perform them with promptitude and precision; these are changes of front of a whole line, &c.; the study of minor tactics is therefore principally important as a preparation for partial engagements, for combats. In more extensive affairs the movements are more simple, and are little else than deployments and formations in column, the passages of lines, the formations of squares, of echelons, &c., which are in common use. But our object now is not to discuss or explain tactical manœuvres which are prescribed by regulations; it is taken for granted that these are well known.

A commander who expects an engagement should cause his troops to take some food, for the reason that it can no longer be done after the action begins, and men who are fasting cannot be expected to display much vigor. If the weather is cold, they should not be marched across a ford to meet the enemy, because half-frozen limbs are not active, and a soldier who is thoroughly chilled is already half beaten. An exception may be made to this rule in case of passing through the water to use the bayonet, because the blood is then kept warm by the excitement of the moment, and such a movement will generally succeed on account of its daring.

Before engaging, an inspection of the arms should be made, to see that all are in good condition, that nothing is wanting, the cartridge-boxes full, &c. The general should assemble the principal subordinates and inform them of his intentions and expectations, as well as all he knows of the plans of the enemy. He encourages them by the confidence thus manifested, and a proper spirit of emulation is excited. He especially recommends them to lend mutual assistance, to avoid partial efforts, to act impetuously and together, to show a fine example to the soldiers, and to do nothing but what is perfectly honorable. Points of assembly in case of reverse are agreed upon, and they are then sent back to their respective commands.

If the combat is foreseen the troops should fight in full uniform. This mark of politeness is due to a respectable enemy, and troops who are trying to appear to the best advantage will usually do their duty best in the engagement.

In a Plain.—The action is begun by the skirmishers, who are spread out to the front sufficiently far to cover the deployments; and when, after a longer or shorter time, according to circumstances, the skirmishers have unmasked the front and rallied to the rear, the firing of the real line will begin. The firing should not be commenced at too great a distance, for, in that case, some of the shots are lost; neither, from an excess of confidence, which has sometimes proved fatal, should the firing be delayed until the enemy gets too near, because the ranks, having to receive several discharges, become somewhat thinned, especially as the first shots are always better aimed than the subsequent ones. Firing by file is almost exclusively used, because that by platoons can seldom be maintained, on account of the noise and confusion; and, moreover, the soldier fires more accurately when he takes his own time. Volley firing, by battalion or half battalion, sometimes has a good effect, as, for example, when poured into a body of troops boldly advancing to the charge. In such a case as this, boldness should be met by boldness, the fire reserved until the enemy is within thirty or forty paces, a volley delivered in his face, and then a rush made upon him with the bayonet. A body of troops advancing find themselves but little injured by distant firing, their courage is really increased, and they become irresistible. If, on the contrary, the fire is reserved, the men are disconcerted at their unexpected reception, and when at last the volley comes, they are ready to take to their heels. If this murderous discharge is followed up by a rapid charge, it will almost certainly prove successful.

While a fusilade is going on from both sides, one or the other will insensibly gain ground, by that instinct of the soldier which urges him to press upon an adversary he deems his inferior. Thus, without an apparent movement, without any command being given, a wing will be gradually advanced. Then a reserve force should be brought up and deployed against the yielding enemy, or thrown upon him in column. If this attack is successful, the whole line advances and attacks the enemy with the bayonet, marching in line of battle, or forming columns of attack with skirmishers in the intervals.

This last course is preferable in ordinary cases; the commanders encourage their columns, and correct the disorder which is almost inevitable in those which have suffered most. As soon as a battalion has broken the line of the enemy, it should prepare to attack in flank the nearest troops which are still in good order. For that purpose, each column should be so organized that it may readily be divided in two parts, and moved by both flanks; the column of attack permits this. If the enemy's line is broken in this way at two or three places, it cannot stand much longer at any point.

But when the opposing bodies are very near, it is sometimes better to charge in line than to take time to form the columns of attack. If the battalions are well drilled, this last method is very effectual in breaking the line of the enemy when he seems to be wavering, or has suffered greatly. By attacking thus along the whole front, the enemy is unable to practise the stratagem of making an opening in his line, with the purpose of drawing on the assailants, and then enveloping and destroying them.

Sometimes the line will advance upon the enemy with some battalions deployed, and others in columns, the latter being formed under cover of the skirmishers, and the deployed battalions continuing their fire. The whole line then moves forward, the deployed battalions halt at short range, and pour in their fire, while the columns rush upon the enemy, break through his line and threaten the flanks. The deployed battalions advance again and sweep the ground with their fire. Mutual assistance and support are thus rendered by the deployed battalions and those in column. This order is sometimes adopted at the outset of the engagement, to give more solidity to the line, while preserving the advantage of a good front for firing. It will be used when it is desired to act offensively upon a particular part of the enemy's line, to seize the key-point of the field, &c. At the battle of Fuentes de Onore, in 1811, a brigade, composed of five battalions, had three formed in close columns by division, and the two intervening deployed. In his admirable Italian campaign, in passing a ford of the Tagliamento, Bonaparte had in each regiment one battalion deployed, and the other two in close columns on the wings of the first. Oftener still this order of attack results from the fact that, at the moment of advancing, some battalions are formed in column, while others, being better situated for delivering their fire, remain in line. The openings thus formed in the line should be filled by skirmishers.

The rule for charges is, to persist in them when once begun, not to fire a shot, but to press upon the enemy as rapidly as possible, both to avoid the effect of his fire and to overthrow him by the shock. The bayonet should not be brought down until within ten paces of the enemy, as order in the ranks is much better preserved in this way. There is nothing more imposing than a column advancing in this way at the "double quick;" the ground trembles under the tread of many feet.

When the line of the enemy is routed, a few companies should be detached in pursuit, and the ranks are reformed for a renewal of the contest, if necessary, with the second line or the reserve. No longer halt, however, should be permitted than is necessary to rally the troops; an advance should at once be made upon the second line, which, being discouraged by the defeat of the first, and perhaps thrown into confusion by its disordered and flying battalions, will probably make but a faint resistance.

If the combat is obstinate, and his first line has suffered severely, the general should bring up his second, and move it through the intervals of the first. This offensive movement, made under cover of the skirmishers, will shake the enemy's line; he will with difficulty be able to stand the fire of these new battalions, that have just deployed their fresh troops in front of his, who are wearied and harassed. Here it should be observed, that if the passage of the lines were effected, by the battalions of the first line falling back through the intervals of the second, the result would have been much more doubtful. Every retrograde movement is dangerous, because of its bad effect upon the *morale* of the troops. To fall back is, with most men, to acknowledge themselves worsted. It is particularly important to avoid movements to the rear with raw troops, who easily lose their equanimity.

In the mean time, the first line becomes the second, reforms, sends off its wounded men, takes some rest, and makes preparations for new efforts, in order to bring the contest to a close.

But fortune may be adverse, and the general will then observe his troops begin to give ground, and to waver and fall into confusion. A portion of the reserve should then be brought up, and if these fresh troops do not succeed in regaining the lost ground, it is time to prepare for retreat. The first line will already, and in spite of its efforts, have fallen back towards the second line, and may be so near, that, to avoid producing disorder in the latter, it may be necessary to effect a passage of the first to the rear of the second. If the first line falls back in good order, it should halt at three or four hundred paces to the rear, and face about; but if some of the battalions are scattered, their chiefs should plant their colors in some conspicuous places, and cause the drums to beat at the same time. The soldiers will have rallying points in this way.

In the mean time, the second line cannot hold its ground long against the victorious troops of the enemy; it must therefore commence to retire in an orderly way, either falling back in line, occasionally halting, turning upon the enemy and giving him several volleys; or adopting the checker formation, either by battalions or half battalions, or moving in echelon, if one wing is less pressed than the other. At the same time, skirmishers should, if practicable, be thrown out upon the flanks of the enemy, to check his pursuit. This is the moment for engaging the troops which have thus far been held in reserve. These will pass to the front, and rush impetuously upon the enemy with the bayonet; by vigorous conduct of this sort they may check his advance, or oblige him at least to move with more slowness and circumspection. In this way efforts will be made to dispute the field, inch by inch, until night. The baggage and the wounded are sent to the rear, to be out of the way of the retreating troops. If there is a defile, such as a bridge, or causeway through a marsh, &c., to be passed, the troops must form line of battle in front of it, and perform the manœuvres prescribed for passing defiles in retreat. If the defile is short, the troops will reform on the other side; if it is of considerable length, and the enemy presses on in pursuit, a defence must be made from point to point, advantage being taken of the ground. Sometimes an ambuscade may be prepared for an incautious enemy. This is a resource upon which too much reliance should not be placed. Still it should not be forgotten or neglected.

It may happen, and, unfortunately, often does happen, that a vigorous defence of a position against superior numbers may result in the defenders being surrounded. Such a state of affairs is certainly very critical, but it need not necessarily end in a surrender. On the contrary, the commander should gather all his forces, before he is too closely pressed, and rush headlong upon the enemy at the point where his line seems weakest, or in the direction where the chances of escape are greatest. Fortune usually smiles upon such efforts; the enveloping line is weakened by its extent, is pierced, and all or part of the troops escape. In 1795, the Piedmontese general, Roccaviva, being surrounded at Loano by a French division, refused to surrender, fell in mass upon the enveloping line, broke through it, and succeeded in joining the Austrian army. Even if such an attempt fails, the captured soldiers will at least have done every thing required by the most exalted honor.

Upon Heights.—Combats do not always occur in level ground; on the contrary, undulations of the ground are often advantageous to the defence, and lead to its occupation. It then becomes necessary to attack the enemy in position, if he cannot be turned.

A height may be attacked either in front or on the side. Most frequently both attacks are used together, because it is necessary to divide the attention of the enemy. Without that, he might have too much the advantage. If the defenders are uncovered, occupying the slopes and not the crest, the attack may be opened by firing upon him, to thin his ranks. But this will not be kept up long, as the advancing columns will soon put an end to it; they will slowly ascend the slopes, stopping occasionally to take breath. When they have nearly reached the enemy, they should increase the pace, and charge with the bayonet. The columns should be rather numerous than long; too great length makes them heavy, and they present too good a mark for the fire of the enemy. They should all advance equally, otherwise those leading may be seriously checked, and the effect produced is bad.

If the enemy is entirely on the summit, it will be useless to fire, as the crest hides him, unless he has some troops on the slopes for the purpose of firing. The columns, being in readiness to attack after ascending the slopes, halt to take breath and reform, especially if, from the form of the ground and the retired position of the enemy, both parties are concealed from each other. A warm reception and a murderous fire must be expected, for the defenders, being three-quarters covered by the ground, have a very great advantage over the assailants when they first show themselves. Moreover, the defenders are fresh and in line, whereas the assailants are more or less fatigued and in column. Therefore, when the movement is begun, it should be made with great impetuosity and rapidity, to prevent the defenders from firing even a single volley, if possible. A ruse may sometimes be resorted to, to draw their fire, by sending forward a few skirmishers in open order, and, under their protection, the main body may advance, the men stooping down and raising their caps on the end of their bayonets, so as to make the fire pass over their heads; if the enemy fires, an immediate rush must be made upon him.

The columns should always be accompanied by numerous skirmishers, closing the intervals between them and covering the flanks. These skirmishers, forming a closer line than usual, fire rapidly; they act like a deployed line, preparing the way by its fire for the advance of the columns of attack.

The course to be pursued in defending heights is indicated by what has already been said. The troops who first show themselves having been received by a well-aimed fire, a rush should be made upon them without reloading; when they are repulsed, the defenders may return to their former covered position. This method was more than once successfully practised by the English against the attacks of the French columns, and particularly at Pampeluna, in 1813. If the defenders do not halt and fall back to their position, after repelling the enemy, they run the risk of being drawn on by a pretended flight to the bottom of the hill, where the assailants turn upon them and defeat them. It is never safe to indulge in a headlong pursuit, even of a disordered enemy. Certain precautions should always be taken, which have been more than once pointed out.

If there are fortifications on the high ground, care should be taken not to mask their fire, as otherwise their whole advantage may be neutralized. General Taupin committed a fault of this kind at the battle of Toulouse, where he placed his troops in front of a redoubt, which was really the strong point of his position. He made another mistake which it may be well to point out, as it relates to the subject now under consideration. He formed his whole division into a single column, which was enveloped by the fire of the enemy, and could only reply from the head of the leading battalion; it lost its momentum, was repulsed, and driven in disorder off the ground. This kind of attack was proper against an enemy with a stream behind him, and climbing a hill with difficulty, but there should have been several columns, and not a single one. If the French had been less impetuous, and had been deployed upon the top of the high ground, imitating the method used by their adversaries, the presumption is a reasonable one that they might have driven them back to the stream and into it.

Hence it appears that, if the defenders are deployed, and the enemy advance in a single column, there is no cause for apprehension. The line should be formed in a circle, or like the letter V, in order to envelop the column, which can only fire from its head. Coolness and good aiming are all that are requisite, and this formidable column will almost surely be checked in disorder; if it continues to move on, resistance to it in front is useless; the line should yield and the attack be made upon the flanks; not a single shot will be lost, and the overthrow of the enemy is certain.

It is more difficult to resist the attack of several columns. However, if it is foreseen, if the troops on the defensive are in good order, a steady fire should be opened at good range. The flank companies will deliver their fire upon the flanks of the advancing columns, and the enemy, seeing his losses increase, will either halt in order to deploy and return the fire, or his impetuosity will at least be much diminished.

In Woods.—If the enemy is holding a wood, skirmishers must be principally used to dislodge him, as great loss would be caused the assailants were they to attack at the outset either in line or in column. The skirmishers should envelop the salient portions of the wood, for they will in this way be enabled to take in flank those men who are trying to conceal themselves behind trees. If there is any part of the wood commanded by neighboring heights, or that may be approached unobserved, the skirmishers should select these points, as they are evidently weak.

In order to get near the point of attack, the assailants should take advantage of all local features, creeping along hedges and ditches, and covering themselves by every little elevation of ground, by clumps of trees, &c. The enemy should be fired upon from every point that is gained, in order to harass him as much as possible; and even when he is really not much injured, the accuracy of his fire will be affected. While efforts are made to advance in front by quite open lines of skirmishers, the attention of the enemy should be diverted by false attacks at distant points.

As soon as the skirmishers have taken possession of the outskirts of the wood and are covered by the trees, detachments should be brought up to their support, and as they penetrate into the forest, the main body advances, divided into several small columns that move forward to the sound of drums and trumpets; they should always be at some distance behind the skirmishers, but ready to support them if necessary. If an open space is found, the troops must be rallied before crossing it, and new measures taken to attack the other part of the forest; the same course should be pursued whenever any obstacle is encountered; no individuals or detached bodies should alone cross a ditch or a ravine or a strong hedge, for there is always danger of meeting the enemy in force on the other side, and it may be impossible to assemble men enough to resist him.

When the forest is of small extent, it is better to try to turn it; in this case, the assailants will keep out of range of fire-arms.

In the defence, an effort should be made to occupy the skirts by skirmishers; if time permits, the adjoining ground should be cleared up, ditches filled, &c., in order to afford a better view of the assailants and to remove every thing which might cover his advance; abatis should be constructed in the most accessible places, and especially by cutting down the trees at the most salient positions; salients may also be flanked by small field-works, in order to prevent the assailant from enveloping them; trunks of trees, laid one upon the other, and held in place by stakes, are excellent for this purpose.

Behind the line of skirmishers, supports are placed, either to secure their retreat or to re-enforce them; they are also on the wings, to prevent them from being turned. Still farther to the rear, and at a point equally distant from the threatened points, is placed the reserve, to act according to circumstances. Thus, for example, if the attack is disconnected and the defence of the outskirts is successful, the reserve may take the offensive by debouching from the wood or by passing around it. Such an operation may succeed, because the troops that execute it are concealed during the movement and the assailant may be surprised.

In Inhabited Places.—The attack of a body of infantry, posted in a town or village which is favorable to defence, is a very difficult operation without the aid of artillery. In such affairs the artillery is the most important arm. However, if the enemy has not had time to fortify himself, the attempt may be made with infantry alone, if the assailant is in superior numbers. We will suppose that infantry alone is used. Generally the attack will be made in columns; skirmishers should envelop the village, to drive the defenders from hedges and garden walls, to fire into windows, and attempt to take possession of several detached houses, where they may take post and fight the enemy on equal terms.

The columns will not advance until the fire of the skirmishers has produced its effect, which will be inferred from the slackening of the fire of the defenders. Until this time the columns should keep out of range or covered by some accident of the ground. While one column attacks in front and attempts to penetrate the principal street of the village, others should be at its sides to look for other entrances, or to turn the place, if possible. The columns will be preceded by sappers or workmen, provided with axes and crowbars, shovels and picks, who throw down walls, cut hedges and palisades, fill up ditches; in a word, remove all obstacles that might impede the movements of the troops. If a house offers a particularly obstinate resistance, it should be surrounded and attacked on all sides; and, if the defenders obstinately refuse to lay down their arms, they should be threatened with having the house burned over their heads. For that purpose straw and piles of brush will be brought up, but fire should not be set to it until the defenders have again been summoned to surrender, as no unnecessary cruelty should be practised even in war.

If the enemy defends the village from house to house, the same course must be pursued by the assailants, that is to say, fire will be poured from the windows and roofs of those which are in possession of the latter. It will readily be perceived that this kind of a contest can only take place in large villages or towns where the streets are, like those of cities, lined with continuous rows of houses solidly constructed. In such places one precaution must be taken: it consists in advancing singly along a street that is to be seized; men slipping along the walls are covered by very small projections, and may succeed at last in reaching important points without too much exposure. Generally, when troops are making an assault in a defile, they should not be crowded, as a repulse of those in front may lead to terrible slaughter and confusion. The advance should, in such cases, be made by successive companies at considerable intervals, those not actually engaged being kept under cover. In proceeding along a street an entrance may be effected in one house, and the assailants may then sometimes pass from house to house under cover, breaking down division walls.

In order to defend a village, the infantry should be distributed in the gardens, behind walls and hedges, and at the windows of houses. A single rank is sufficient for such purposes, and, consequently, a quite extended line may be occupied; however, it should not be more so than necessary. This is a fine opportunity for riflemen and sharpshooters to show their skill, because they are out of sight of the enemy and can quietly take perfect aim. Platoons or companies should be placed in the streets or avenues, to support the line of sharpshooters, and a reserve at the centre of the village or a little to the rear, in readiness to move in any direction to meet the enemy. If time permits, openings should be pierced in the walls to serve as loopholes; the principal streets should be barricaded, &c.

When the village is surrounded or preceded by a ravine or a little stream, or any other obstacle, the defence should commence there; if necessary, a retreat will be made upon the village, which will have been previously occupied by a portion of the troops and put rapidly in a defensible condition. The resistance that may be made under cover of the houses, walls, and other enclosures, will be the more energetic and prolonged, as the enemy will be less likely to turn the position, either on account of its natural features or the disposition of the troops. In this case, vigorous sorties will succeed well, if made at suitable places and not pushed too far, for fear of the parties being cut off; the avenues by which they will return should be suitably occupied, so that the enemy may not re-enter pellmell with them. Offensive returns of this sort always produce a good effect, and they should be frequently repeated ; they elevate the *morale* of the defenders and disconcert the assailants.

In order to avoid confusion, each detachment should be charged with the defence of a particular part of the village, and instructed in the manner of doing this, and of effecting a retreat, in case such an operation becomes necessary. This precept is applicable not only to the defence of a village, but to any military operation.

After all that has been said, it must be remembered that wooden villages are rather dangerous than useful to occupy, because they are readily set on fire. A village which can be surrounded on all sides cannot be well defended. To make a good defence, it is therefore necessary : 1st, that the village be built of masonry; 2d, that it rest upon some obstacle which prevents its being turned, or else it must be supported in rear by good troops.

Art. II.—Cavalry against Cavalry.

The fundamental rule for cavalry is, not to wait to receive an attack, but to act always offensively, and to be galloping at the moment of meeting the enemy.

Otherwise it will be thrown into confusion and dispersed, for it has been clearly demonstrated by experience that a squadron, even of heavy cavalry, cannot withstand, at a halt, the attack of another squadron of which the horses are very small, provided the latter is at the gallop. This is not because the shock of a body of cavalry is proportional to the mass and the velocity with which it moves, but because the gallop gives impetuosity, excites the horses, and timid riders are drawn along with the others.

When two bodies of cavalry are advancing towards each other to engage in front, that one will have the advantage which has several squadrons in a position to charge at the same time upon the flank of the other. The result of such a movement is even more decisive than in combats of infantry, both on account of the rapidity with which it is executed, and the difficulty the assailed line has in opposing it. There should, therefore, be placed in rear of the wings of a line of cavalry, columns composed of several squadrons or platoons, according to the strength of the force in line. These columns, formed at full distance, while they enable an attack to be made upon the enemy's flank, are the surest means of guarding against a similar attack from him, since a column at full distance may in a moment deploy and be ready to meet an attempt to envelop the extremity of the line. In an attack made by heavy cavalry, the columns on the wings may be formed of light horse, which, in addition to the duty already assigned them, may pursue the enemy after the charge, while the line of battle is reforming. As the enemy may adopt the same measures, the precaution should be taken of holding a small reserve, if only a platoon, in echelon with the columns on the wings. To avoid inversions in deploying these columns, that on the right wing should have its left in front, and that on the left its right in front. But, as has already been explained, cavalry should be perfectly accustomed to inversions, in order to be always ready to deploy in either direction.

When cavalry can rest its wings on any natural obstacle which prevents the enemy from manœuvring on the flank, the columns on the wings are no longer necessary, but, as a general rule, their presence cannot be too strongly insisted on. It appears, therefore, that a body of cavalry moving to the attack, is partly deployed and partly in column. While satisfying these conditions, it should occupy as large a front as possible, because it is important to have as many engaged at once as possible. If a body of cavalry should advance in column, it would be almost certainly beaten, as none but the men in front could use their weapons; enveloped and attacked on the flanks, the column could only extricate itself by a speedy flight.

If cavalry advances in a single unbroken line, with but small intervals between the squadrons, it is said to charge *en muraille*, or *as a wall.* This method of attack is very imposing, but is only practicable for a few squadrons, because, on account of the irregularities of the ground and other obstacles, and the inevitable crowding that takes place in a line of considerable length, the shock is not a single one, but composed of partial shocks, whose effect is by no means so great; if the line is thrown into confusion at any point, and especially if it is pierced by the enemy, the disorder may become general throughout the line; mistakes are corrected with difficulty; finally, one of the principal advantages of this arm—its mobility—is lost. The charge in a continued line of great extent is hardly practicable, when the design is to sweep over a field of battle covered with broken battalions, of which some are making a show of resistance.

The attack in echelons is much used by cavalry commanders, and offers the advantages of not engaging the whole force at once, and of giving better opportunities of retrieving chance reverses during the action. The most retired echelons are last engaged, and, up to the moment of being so, are disposable for the support of the others, or to attack the enemy in flank. This formation is especially advantageous when the cavalry, in order to attack, passes from column into line, as would happen, for example, in issuing from a defile; under such circumstances it is not necessary to wait until all the echelons are in line before making the charge, as it is sufficient for the first to be formed, and the others will come up in succession. So long as the deployment is unfinished, the enemy is uncertain where the attack is to be made, and this is an advantage for the assailant. Two or three echelons may be placed in position to threaten his left; he will re-enforce that part. The remaining echelons may then be launched against his weakened right, which will very probably give way, as there is not time enough for a counter-manœuvre on his part. If the leading echelon is successful, it may take the line in flank while the others attack in front. The echelons must have considerable strength in themselves, that their successive shocks may produce the expected effect. They should, therefore, be formed of regiments, or, at least, of squadrons; echelons of platoons are of little value. We repeat, cavalry in attacking cavalry should always form a front of considerable extent, without, however, dispensing with the columns on the wings or other means of attacking the enemy in flank at the same time as in front.

Whatever dispositions are adopted, the rules for the charge are invariable. The troops move off at a walk, then taking the trot, which becomes gradually more rapid until they fall into the gallop at some distance from the enemy, and at last into the full run when at about a hundred paces, uttering at the same time loud cries. Without these precautions, if the gallop is taken too soon, the horses are blown and the ranks disunited before the enemy is reached, the imposing effect of the onset of one great mass is lost, the shocks are partial and of no avail. The shouts inspire the men and excite the horses. At the moment of the shock and in the ensuing mêlée, it is better generally to use the point rather than the edge of the sabre.

If light cavalry finds itself in presence of heavy cavalry, it should not receive attack, but should scatter and charge in dispersed order upon the flanks, sometimes cantering around the line, flying before it, and eluding it by superior speed and agility, sometimes pouring in a fire from musketoons or pistols. This is almost the only case where fire-arms can be employed in combats of cavalry, for the musketoon is placed in the hands of the horsemen, not for use in line of battle, but simply for skirmishing, for getting out of a difficult position when dismounted, and for other exceptional cases. To stop during a charge and fire would be a great mistake, as the momentum would be lost, and the attack almost certainly repelled. It would be equally an error to receive an attack in this way, for the line would be routed before the musketoons could be put away and the sabres be drawn.

When the cavalry is numerous, it is formed in two lines, like infantry. The first is always deployed. The second, which is often inferior in numbers, is either deployed in part and outflanking the wings of the first, so as to prevent its being turned, or formed in as many columns of platoons at full distance as there are squadrons. All these columns are ready to form line of battle by a very rapid movement to the front, and they also leave wide intervals between them, through which the squadrons of the first line may readily pass to the rear after being routed or thrown into confusion by an unsuccessful charge. Without this precaution, the second line would run the risk of being pressed back by the first. The second line remains at several hundred paces from the first, whose movements it follows, whether in advance or retreat. If the first line is repulsed, the second sends its flank squadrons to take the enemy in flank, and disengage the first. In the mean time, the columns move to the front at a trot, in order to deploy as soon as they are unmasked, and to rush upon the enemy's line, which must be also more or less in confusion. Thus, in combats of cavalry, a body, at first victorious, is checked by other squadrons taking the places of those that have been beaten, and the whole face of affairs is changed in a moment.

To avoid accidents, it is well *always* to rally, even after the most brilliant success. But it is not necessary that all should halt: several platoons in dispersed order should harass the enemy, while the standards are moved forward at a walk. The soldiers, obedient to the call of the trumpet, will regain their places in ranks, and the squadrons, all the time advancing, are soon reformed and ready to charge again.

The close column is best for the manœuvres of infantry on the field of battle; but, for cavalry, the usual formation is at full distance, because the subdivisions may thus with greater facility and rapidity form line in whatever direction the enemy may appear. However, when cavalry has nothing to fear upon its flanks, it may manœuvre at half distance, in order to have less depth of column. The enemy may be deceived as to the number of cavalry, if some columns move at half and others at full distance.

Art. III.—Cavalry against Infantry.

Charges against infantry are principally made in columns of squadrons at double distance, so that, if one squadron is repulsed, the men may turn to the right and left and rally behind the column. The second squadron charges and performs the same manœuvre as the first; the third comes in turn, and then the fourth. If the infantry is not very firm, and does not manage its fire well, it will undoubtedly be broken.

Attacks in columns at double distance are chiefly directed against the angles of squares which are not well defended. Against deployed lines, that cannot be turned, charges are usually made in echelon. However, charges in column would answer well in such a case, if made simultaneously at several points. Both are better than a charge in a continuous line against infantry, because the loss of a squadron does not check the whole, and there is sufficient freedom of movement to allow attack upon weak spots, or those temporarily deprived of fire; finally, renewed or successive shocks are more exhausting to infantry, and produce more real effect than a single charge, however grand and imposing the latter may be. A charge in an unbroken line against infantry can only succeed when the latter has been much cut up and disorganized by artillery, or when bad weather prevents them from firing. At the battle of Dresden, Murat took advantage of such a state of affairs, broke a line of Austrian infantry, and sabred many of the men.

Cavalry should avoid passing near woods and thickets, unless certain they do not conceal infantry skirmishers, as much injury might be caused them by the fire of the infantry, to which no reply could be made.

When a charge is successful, and the enemy's line broken, the important thing is not to sabre the fugitives, but to turn upon those who still stand their ground. For this purpose, the cavalry is rallied, and manœuvred to envelop the flank of the portions of the line yet fighting. The infantry will be almost certainly swept away.

When, on the contrary, a charge is repulsed, the horsemen should gallop back to the place for rallying. If there is a second line, it should be promptly unmasked, to give it an opportunity of coming into action. The ranks are reformed, preparations are made for a new charge, as the failure of the first attempt should be no cause for discouragement. Good cavalry will charge again and again.

When infantry is formed in square it should dismiss all apprehensions, and pour into the cavalry a well-aimed fire, that should not be begun too soon; volleys, commenced when the cavalry is at about one hundred yards, will have a good effect. The fire by ranks is the best to use. In this way, four well-aimed volleys may be delivered, and without precipitation, while the cavalry is passing over the one hundred yards which separates it from the infantry when the firing begins. This supposes the front of the square to consist of four ranks. First, the two rearmost ranks fire successively, while the men of the two foremost ranks stoop. These then rise and fire at the word of command, the front rank last, and in the very faces of the cavalry. In the mean time, the two rear ranks have reloaded their arms, and may again use them against those horsemen who may have succeeded in reaching the line. This firing by ranks, when there are four, is only practicable when each rank fires but once, for a renewal of it would be dangerous to the men of the front ranks, in the midst of the smoke and confusion of the combat. Four successive volleys, well aimed, are enough to check the best cavalry. Four ranks are absolutely necessary, unless artillery is at hand to keep the cavalry at a distance. Infantry, formed in two ranks, is too weak for such combats; at least three are necessary, and four are better.

The bayonet is the last resource against cavalry that has not been checked by fire. It has been proposed to stretch ropes, and place chevaux-de-frise in front of the infantry, but these devices, which answer very well on a drill-ground, are of no practical value in presence of an enemy; they simply embarrass the infantry, and clog its movements. The foot-soldier may and should expect to repel cavalry by the use of his ordinary weapon; he should be cool, and convinced that horses can never pass through a line of infantry well closed up and bristling with bayonets; timidity, confusion, and the ravages of artillery are the only things that can compromise him. Let alarm be felt neither at the approach of the horses, nor the cries of the horsemen; let the ranks be well closed up; let the fire be carefully husbanded, and there need be no doubt as to the result of a charge.

It is not to be expected that this confidence in their weapons will be felt by other than veterans, who have had occasion to test their value, and know, from experience, what they are worth. It is for such as these to ask, as was done in Egypt and at Marengo, that the cavalry be permitted to come within fifty paces, and even less, when a most destructive fire may be poured into their ranks. It would be dangerous to try this with raw troops, who may experience a not unreasonable feeling of apprehension at the approach of cavalry upon them for the first time. For such soldiers, one hundred paces is about the distance at which they should begin to fire.

The resistance of infantry to cavalry is chiefly due to the order of battle adopted for that purpose. Its power of resistance is greatly increased by forming several squares, which protect each other by their fire; and the flank and reverse fire, procured in this way, is much more feared by the cavalry than that coming in front. Mention has been made in another place of large squares, formed by divisions or brigades. We have now reference to battalion squares, because we are considering a simple combat between a body of infantry and one of cavalry; in this case, the smallness of the squares is not only not disadvantageous, as interior space is not required, but several small squares are better than a single large one, because, if the latter is broken through on one front, all is lost; but, one or more of a combination of small squares may be destroyed, and the remainder hold firm; besides, they give mutual support by cross and flank fires, and the enemy's attention is, moreover, divided.

The formation of these battalion squares is very simple, and is explained in detail in the authorized works on tactics. There may be either a line of oblique squares, or in echelon, with a view of flanking each other. Squares, if small, may be moved considerable distances without material inconvenience.

The infantry should carefully avoid throwing away its fire upon skirmishers sent forward by the cavalry, often for the very purpose of drawing it, or raising a dust, under cover of which some movement may be concealed. If the infantry imprudently delivers its fire upon these skirmishers, the squadrons should immediately charge, before the arms can be reloaded. The skirmishers should be kept at a distance, by sending out a few good marksmen in front of the squares.

The only manœuvres and formations infantry may safely adopt in presence of cavalry, are those of squares and close columns.

Art. IV.—Combats of Artillery.

Artillery should never be brought into action, unless well supported by other troops. The supports take positions to the right and left of the batteries, covering themselves, by any features of the ground, from the fire of the enemy, only emerging to resist attacks upon the battery or upon themselves. Positions are selected for artillery where it may act with greatest effect. A firm soil is requisite to prevent the wheels from sinking too much. A rocky site would be unfavorable, from the many dangerous fragments produced by the shot of the enemy. The space should be sufficiently extensive to afford room for all the pieces of the battery, and high enough to give a good command over the surrounding ground within range, but not so high that the fire would be too plunging. Slopes of about one-hundredth are the best. If the object aimed at is near, the elevation of the site of the battery should be small, but this may be greater as the range increases.

Artillery duels rarely lead to decisive results, and are usually avoided. Sometimes they are inevitable. In such a case, other things being equal, that party will evidently have the advantage which has the best view of the enemy, and has at the same time the best cover. Artillery cannot endure a flank or very oblique fire, and, when such a fire is received, an immediate change of position becomes necessary. This may sometimes be accomplished under cover of the smoke, or some undulation of the ground. At the battle of Iéna, Colonel Seruzier was in command of a battery of twenty pieces, which was opposed by a very superior one, and his was on the point of being destroyed. He immediately withdrew the alternate pieces, keeping up the fire with the others, until, by a rapid movement under cover of the smoke, he reached a position where he had a very slant fire upon the enemy; his remaining pieces were soon disengaged.

A battery, in taking position to get an enfilading or slant fire upon an opposing battery, runs the risk of exposing its own flank to attack. A battery may present a poor mark to the enemy by having wide intervals between its pieces, while, at the same time, some of them must obtain a slant fire upon him; the fire of all the pieces is convergent. The effect of such a fire is very powerful. A battery of eight pieces united would scarcely be able to contend successfully with one of six pieces divided into two or three parts, at some distance from each other, but all converging their fire on the larger battery. It should be observed that in this we do not find a violation of the principle which sanctions the employment of as many pieces as possible in a battery. Concentration of fire is the object to be attained in both cases, and there is no contradiction between the two.

The caissons should be covered as much as possible from fire, and even from sight. It is better to aim low than high, to use the ricochet when it is practicable, to fire slowly at distances greater than 700 yards, and rapidly inside of that, and every shot should be well aimed. The use of shells is recommended for broken ground; solid shot against columns of attack, although, at close range, grape and canister are better, as also against cavalry and infantry in line. A strict defensive should not be observed, for if it is well to select a strong position, and to profit by every accidental cover presented by the ground, it is equally so to inspire the enemy with an apprehension of attack, and to advance upon his position when prudence permits, as the moral and real effect of such a course is great.

The cavalry is the best arm for supporting that portion of the artillery which should move rapidly from point to point, but the supports of batteries of position should be of infantry, and, indeed, that arm is the best in all cases when the simple protection of the artillery is considered. If both arms are united as supports, infantry skirmishers will be to the right and left to engage the opposing skirmishers; the main body of the infantry will be somewhat to the rear, as well covered as possible, and the cavalry in echelon on the flanks.

A battery, whose supports are very weak or do not contain cavalry to meet that of the enemy, may be in danger of being captured, and should select a position unfavorable to the action of cavalry. Colonel Foy, seeing a body of Russian cavalry issuing from the bridge near Schaffhause, rapidly withdrew his artillery into a clump of pines which was fortunately at hand; he there formed his battery into a square, and was enabled to keep the cavalry off until two regiments of infantry came up to his assistance.

When a battery is very destructive in its effects, it may become necessary to attempt its capture. Skirmishers will precede the assaulting party, the intervals of the line being as great as possible, to offer a poor mark to the enemy, and to enable the men to shelter themselves behind any little accidental covers. These skirmishers will be specially useful, if they are good marksmen, in killing the gunners and horses. While skirmishers advance on the front of the battery, small columns will threaten the flanks. If the fire is directed upon the columns, the skirmishers will advance rapidly upon the front; if, on the contrary, the skirmishers attract the fire, the columns may advance without much loss. When these are near the supports, they will charge vigorously, and if the supports are driven off, the battery is attacked in rear.

As cavalry moves very rapidly, it is a good arrangement to have the assaulting party contain both infantry and cavalry, the former to act as skirmishers and occupy the battery in front, while the latter form the columns on the flanks.

If the pieces are captured, and the horses are still at hand for use, they should be rapidly carried off; if this is impracticable, they should be spiked, and the elevating screws and rammers carried off.

Art. V.—Attack and Defence of a Redoubt.

A redoubt may be armed with cannon, or defended by infantry alone. In the first case, the cannon must be silenced by other cannon; but in the second case, the attack may be made without any such preliminaries.

Marksmen envelop a part of the redoubt, aiming at the crest of the parapet to prevent the defenders from showing themselves, or, at least, to cause them to fire rapidly and inaccurately. The circle of skirmishers is gradually contracted, an active fire being all the time kept up. When sufficiently near the ditch, they advance at a run and jump into it, unless they are checked by some such obstacle as abatis, palisades, trous-de-loup, &c.; if these are encountered, a way must be opened through them before farther progress is made. While some descend into the ditch, others remain upon the counterscarp to fire upon the defenders who show themselves. When the troops in the ditch have rested somewhat, they make the assault. The soldiers assist each other in ascending the scarp and getting upon the berme, from which point they rush over the parapet in a body, and compel the defenders to lay down their arms.

If the redoubt is armed with cannon, and of greater strength than is supposed above, it would be necessary to use artillery, to break down palisades, dismount the pieces, and breach the parapets. The best positions for the artillery are those from which a plunging or enfilading fire can be obtained. If the redoubt has embrasures, one or two pieces should be directed against each of them, as there is an opportunity, not only of dismounting the pieces, but also of firing through the embrasures into the interior of the work. A few good marksmen may also be detailed to pick off the gunners, whenever they show themselves at the embrasures.

When the artillery has produced its full effect, the skirmishers will be advanced, as above described. The skirmishers will generally be followed by supporting columns, that will advance upon the salients of the work. Each column should be preceded by workmen provided with axes, and ladders should be carried by men detailed for the purpose. It is a good plan to give each of the leading men a fascine, with which he may shield his body greatly, and he may use it for partly filling up the ditch. The skirmishers open out and permit the columns to pass through their line, keeping up a rapid fire until the ascent of the parapet by the storming party begins. The important thing, at this decisive moment, is unity of movement and action, an entrance being made on all sides at once. The troops should, therefore, pause a moment on the berme, until a signal is given, and then all will rush forward together. Such an attack as this is almost certain of success.

The commander of a redoubt should use every effort to increase the confidence of his men, especially by his own demeanor; still every thing should be kept in the best possible order. If the attack is not to take place at once, the redoubt should be surrounded with an abatis; large stones should be obtained to roll down into the ditches; sand-bags and sods should be prepared for making loopholes along the parapet. As soon as the batteries of the enemy are seen, fire should be at once opened upon them; but when their guns are fairly in position and partly covered by the ground, the contest becomes an unequal one. Only those pieces will be kept in position which are covered from the enemy's fire by traverses or otherwise. A few other pieces may sometimes be temporarily run into battery and fired with grape upon the infantry. The effect will be good.

A few marksmen will at first shelter themselves wherever they can along the parapet; but when the fire of the assailants ceases, for fear of injuring their own men, the defenders will line the parapets and open a warm fire of musketry and artillery upon the columns of attack, and upon the skirmishers who are attempting to pass the obstacles in front of the ditch. This is the time to explode any mines that may have been prepared in the ground over which the assailants must pass. If the enemy succeed in getting into the ditch, and collecting their forces to make the assault, shells should be rolled down upon them, as well as trunks of trees, large rocks, beehives—any thing that may cause confusion. The defenders must then mount the parapet to receive the enemy at the point of the bayonet. More than one attack has been repelled at this stage of it. At Huningue, in 1796, Captain Foy of the artillery, observing that the enemy had planted scaling ladders to ascend into a demilune he was holding, had shells rolled down upon them, by which considerable injury was done them, crowded as as they were in the ditch. The gunners seized their rammers and handspikes, and, mounting the parapet, knocked on the head and back into the ditch the first men who showed themselves. The enemy abandoned the attack. When we reflect upon the disordered condition of the assailants at such a moment, and the difficulty they must experience in using their weapons, as well as upon the physical and moral advantages in favor of the defenders, who have a firm foothold upon the parapet and nothing to clog their freedom of motion, it will be manifest that, not only have the defenders many chances of success, but it seems impossible that such an attack can result in any thing but failure.

Art. VI.—Attack and Defence of a strong Dwelling-house, with out-buildings and enclosures, all of masonry.

A careful examination of the locality should be made before the attack, and this may be attempted with least danger just before dawn. Nothing is supposed to be known of the arrangement of the interior. All that meets the eye are several walls pierced with loopholes, and several ways of approach along which an attack may be made. By the time the outskirts are seized, there will be light enough to see how to advance farther. Much danger will be incurred in the onward movements, and it might well be desired to have them made under cover of the darkness, but this always leads to confusion and gives those who are so inclined an opportunity to skulk. In night attacks there is generally a want of concert, mistakes occur, panics take place, by all of which means the best plans miscarry. Therefore it is better to make every preparation for the attack in the night and execute it just at dawn. Darkness is only favorable in case of a surprise.

Every preparation should be made for forcing locks and hinges and breaking through doors. If none of the soldiers have the skill necessary for using tools, native mechanics must be pressed into the service, and made to act, if they are unwilling. Ladders should be collected, in order to facilitate entrance by the windows, or, at least, to enable grenades to be thrown in. The attack is supposed to be made by infantry alone, for artillery can soon destroy such structures as we are now considering.

The approach should be made in silence, until the sentinels challenge and the enemy begins to fire; a rush must then be made upon several points at once, in order to divide his attention. A loopholed wall is supposed to be the first obstacle, but by inserting the muskets of several men into each hole, the enemy may soon be forced to leave it. The gate of the enclosure must then be broken down, or a portion of the wall. The foot of the wall of the house is reached; the first thing is to stop up the lower loopholes with bags of earth or with timbers; efforts must be made to break down the door with axes, or by battering, or to burst off the hinges; this will soon be accomplished, unless the doors are strongly braced from behind; in the mean time, sharpshooters should watch the windows and fire whenever a man shows himself. The men should keep dispersed as much as possible. Whenever a window seems abandoned, attempts should be made to enter it. To increase the number of openings into the interior, the wall may be battered down with any heavy timbers that are at hand. Finally, an entrance is effected into the lower story, and the defend-

14*

ers driven above. They may now be summoned to surrender. If they accept, all is over, and the captives should receive the treatment due to brave men. If they decline to surrender, a large fire lighted below will soon bring them to terms.

In the defence, doors will be barricaded and loopholes made in all the stories, taking care to have them as small as possible and those of the lower story so high that the enemy cannot close them nor fire through them. Banquettes, to stand upon, may be made of plank placed on barrels, boxes, &c. The angles are weak points, and an effort must be made to pierce loopholes there. If the roof will burn, it must be taken off and the upper floor covered with earth. The principal beams of the frame of the house should be propped, so as not to fall if the wall is breached; to close the breach, various household articles may be used. Books are quite impenetrable to musket-balls. Bedsacks and pillow-cases filled with earth, also carpets rolled up, resist bullets well. Balconies may be fitted up as machicoulis, &c.

Art. VII.—Attack and Defence of a Village.

In Art. I. of this chapter some observations were made with reference to combats of infantry in towns and villages, but they were supposed to be of an open character. What follows is to be understood as relating to an intrenched village.

Here the artillery plays the principal part, because it would be otherwise impossible to prostrate the obstacles prepared by the defence to aid in holding a post which may be of great importance, either as a bridge-head, or a strong point of a line of battle, or a pivot for manœuvres. The most favorable positions should therefore be selected that are offered by the surroundings of the village, either commanding points, or others from which the principal street may be enfiladed, or the walls breached that give shelter to the enemy. If a redoubt defends the approaches, it should be first attacked, for success at other points will lead to no decisive result so long as that holds out. Several pieces of artillery may be directed against a single embrasure, and their fire directed to others in succession, as fast as the guns are dismounted or withdrawn. Other pieces should enfilade the terre-plein, break down palisades, &c. All this may be done openly, if time presses and the ground favors, but it is more prudent to make a careful examination of the locality in the first place, and, during the night, throw up epaulements of earth to cover the guns in the position selected for them. It is also the part of a wise man, before attempting a decisive movement, to procure ladders and tools of every kind necessary for throwing down walls and parapets.

These arrangements having been made, fire will be opened at dawn. The artillery will make vigorous efforts to produce the effect expected. There will be the greater probability of this as the pieces are of heavier metal. When the breaches begin to open, and some of the artillery of the work is silenced, the skirmishers may be advanced to envelop with their fire as much of the village as they can. The skirmishers should not mask the artillery, which continues its fire, but they will aim at windows, loopholes, at every man who shows himself. They will be followed by workmen, provided with tools for levelling hedges and palisades, filling up ditches, breaking down gates, enlarging breaches, &c.

In the mean time, the skirmishers close in upon the village. The artillery is now silent, except it may throw shells into the town. The skirmishers still advance, the boldest even approaching the breaches and attempting to enter, while the workmen are busy with their tools, preparing the way for the advance of the columns over the outer obstacles: until now they have been out of range, or concealed by the ground, trees, &c. They move out at a given signal towards the breaches and other points indicated; they destroy any barriers which may still exist, leap into the ditches and scale the parapets, break down palisading, burst open gates by means of powder-bags or levers. During the same time the skirmishers between the columns enter by any available openings, leaping over hedges, climbing walls, wherever they do not meet a too formidable resistance. Finally, the exterior defences are carried; the troops, before penetrating farther, ought to rally and make new dispositions for driving the enemy from any houses still in their possession.

If the defence is obstinate, much caution is requisite in pushing forward into the interior. The outer houses should be seized, as they might be occupied by parties of the defenders, who could take the troops in the streets at great disadvantage. If the houses are detached, they will be captured by surrounding them; if contiguous and vigorously defended, a passage may be effected from one to another in succession by breaking through the separating walls. The defenders having thus been driven from point to point, an effort will be made to cut them off from the line of retreat, which is usually through a defile, such as a bridge or narrow street; this manœuvre will generally hasten the evacuation of the village.

A village is seldom without support from troops on the exterior; it would, therefore, be impossible to attempt an attack, without having force enough to hold these supports in check or to engage them. To obtain success a marked superiority of forces is requisite, as well as in the number and calibre of cannon. It would be very rash to attack an intrenched village, which formed part of a line of battle, without this superiority. An isolated village should be turned, rather than attacked.

Some time is necessary to prepare for the defence of a village, for there is much to be done if a vigorous resistance is to be made. A continuous enclosure must be arranged, advantage being taken of hedges and walls which will answer the purpose, and all others being thrown down, in order to give no covers for the enemy. Those may be left standing which are completely under fire, and will be hindrances to their movements. The same course will be pursued with ditches. Behind the enclosure, every facility for circulation of troops should be found. The enclosure should be a broken line in order that its parts may flank each other and give cross-fires. If it is interrupted at any point, the break should be supplied by abatis, palisades, barricades, or earthworks; batteries are constructed for the cannon, and sometimes redoubts built.

Behind the standing hedges, trenches may be dug and the earth thrown to the front, as shelter from the fire of infantry is thus obtained. A wall that is over six feet high is more valuable; loopholes may be made in it, and they should be small and beyond the reach of the enemy. A banquette for them, of boards, may be supported on barrels or any thing of that kind. It is not always easy to make these loopholes of suitable dimensions, as the stones in the wall may be too large; this defect may be cured by using stiff mortar or sods.

Finally, a keep should be arranged in the interior of the village, to secure the retreat of the defenders, and to give them more confidence in resistance. This may be a church and cemetery, or a large dwelling-house and enclosure.

A fortified village is a strong point of support for

an army on a field of battle, as frequent examples have proved.

Art. VIII.—Observations upon the Subject of Manœuvres.

After what has been said with reference to various operations in war, in which the manœuvres and evolutions prescribed in the tactics find applications, it will not be out of place to close this chapter by some remarks with reference to these movements, in order that their spirit may be appreciated. Their characteristics should be as follows:

1st. *Promptness;* for, while a manœuvre is in progress, the troops are comparatively helpless as a mass, and it should be consummated as speedily as possible.

2d. *Simplicity*, in order that officers and soldiers may execute the manœuvre mechanically, so to speak, and without thinking of it. A complicated movement produces confusion, and this may lead to defeat. Simplicity, in this case, means facility of execution.

3d. *Security;* that is to say, if the enemy appears unexpectedly while the movement is being executed, the troops should be in the best order to resist him.

4th. *Precision*, in order that every movement may have well-defined limits, which regulate its progress and direction.

5th. *A brief and clear command;* for upon the nature of the command often depends the success of a manœuvre. If the command is such, that a man hearing it cannot doubt as to the meaning, the manœuvre will succeed. If, on the contrary, the language is ambiguous in sense, or the words improperly arranged, there can be neither certainty of conception nor promptness of execution.

A manœuvre, to be good, should, as far as possible, be *offensive;* that is to say, it is better in its execution for the troops to move forward than to the rear, because a retrograde movement has a disheartening effect, and may bring on a shameful flight; on the contrary, the courage of the troops is better sustained when they advance, as it were, to meet the enemy. Manœuvres should be not only well arranged, but properly executed. Skill can only be gained and retained by constant practice, and therefore the drill is necessary for veterans as well as for recruits. While keeping soldiers in good drill, it is necessary to avoid insisting too much upon things really of little importance. Insist upon what is necessary, and reject what is superfluous. Soldiers should be accustomed to remain together and in good order, but it is sufficient, with this view, to require the *touch of the elbows;* that is to say, that each man feel his neighbor lightly, and that the whole battalion, facing in the same direction, be on a line nearly straight. No commander should be so foolish as to bring his "guides on the line" in a shower of grape or musket balls, or should listen to the remonstrances of a major or an adjutant who would think all was lost if the battalion were not in a perfectly straight line. And as to those wonderful firings of a whole battalion as one man, that are sometimes esteemed so highly, is any thing like them ever heard in a great battle? Amid the thundering of the artillery, the beating of drums, the cries of the wounded and dying, the excitement of the combatants, are they able to give that close attention to the word of command which alone makes such firing possible? If possible, what value has it? Are more of the enemy killed and wounded than by a rolling fire? Every operation in war is accompanied by more or less disorder, but this, to the eye of the real soldier, is grand and beautiful when it is the result of a noble daring and enthusiasm, and is infinitely preferable to the frigid regularity of indifference or lukewarmness. It is an excellent thing to drill soldiers upon irregular ground, such as they find upon the field of battle.

Good marching is the important thing. A body of troops that marches well is always imposing, and by this alone victory is often attracted, independently of all the advantages procured by that good order which results from fine marching. It is not out of place to repeat again the maxim of Marshal Saxe, that victory resides in the legs of the soldiers. Marching should, therefore, be deemed of great importance, and much time and pains should be devoted to it, first in level drill grounds, then along roads, and finally over irregular ground of all kinds. The cadenced step, lightness of carriage, and the touch of the elbow are the things for each man to learn; and for the whole body, the preservation of distances and direction.

By keeping the step, wavering is avoided, and a whole battalion moves like a single man. If a body of troops is charging another, it is only by all keeping the same step of uniform length that the enemy's line is attacked along the whole front at once. By the *touch of the elbow*, soldiers remain together and preserve the alignment necessary for an engagement. Any pressure, coming from the direction of the guide, should be yielded to gradually, and without crowding. At the same time, the touch towards the guide shall always be lightly kept up. The *preservation of distances* on the march is of the greatest importance. It is a less evil for the distance to be diminished than increased, because it is much worse to have gaps in a line than to have some men crowded to the rear, who may be usefully employed in replacing or caring for the wounded. The preservation of distances is in the hands of the officers and non-commissioned officers, who, taken together, form what is called the *framework* (cadre) of the battalions. It is of very great importance that the *cadre* be well composed. Young soldiers, directed upon the field of battle by good officers and non-commissioned officers, will rival veterans upon the field of battle, as Lutzen and Bautzen testify.

The preservation of direction is quite as important as distance, in order that crowding and confusion in ranks be avoided, and that there be any certainty in the execution of manœuvres. It is important to practise battalions in running, as it is sometimes necessary in order to achieve success, to get over ground very rapidly, and it is frequently advantageous. It usually produces more or less confusion, and the men should learn to avoid that evil as much as they can. There is a thing rarely seen at drills, which is, however, very important; that is, rallying to the colors, on the supposition that a battalion may be dispersed and rally again to re-enter the engagement. Soldiers should be taught how obligatory it is upon them to rally to the flag as long as it floats, and to sacrifice their lives to save it from capture. Honors paid to the flag are not mere empty ceremony; they are intended to inspire the respect due from every soldier to this symbol of valor and devotion. See the veteran uncover himself in presence of the ensign that he has often followed to victory; he is proud, as of his own wounds, of the marks of the enemy's bullets upon it; he venerates those folds which have survived twenty battles. The names inscribed upon it are his titles of nobility; they bring to mind grand and glorious memories; he values more those brilliant and deserved mementoes of his services than every thing else which men usually so much desire.

CHAPTER VII.

RECONNOISSANCES.

RECONNOISSANCES are either *armed* or *topographical.* Both kinds will be treated briefly, although the last have little to do with our general subject.

Art. I.—Armed Reconnoissances.

Armed reconnoissances are made to procure information with reference to the position and force of the enemy. So long as a commander is ignorant upon these points, he can have no confidence in his measures, either for attack or defence; if he is on the march, he must feel his way with great caution; he is in danger of falling into an ambuscade, or of making a blundering offensive movement, or of being suddenly attacked. It is, therefore, necessary to endeavor not only to learn exactly the composition, strength, and position of the different corps in the opposing army, but to be informed of their movements and simple changes of position, in order to divine the intentions of their commanding general and take precautions to thwart his projects. As much information as practicable should be gathered from intercepted communications, spies, deserters, and travellers. This should be classified under various heads, and always kept complete and ready for reference. With these notes and a good map, upon which are indicated the positions of the several corps, by means of numbers or conventional colors or colored pins, a very good idea may be formed of the force and the location of the hostile forces. When the enemy is very near, it is necessary to ascertain every day what changes of position have been made for the purpose either of concentration or withdrawal. Hence arises the necessity for armed reconnoissances, which are often very bloody affairs.

The custom, usually prevailing, of covering camps with chains of posts and sentinels, makes it impossible generally to see them and the arrangements of the enemy for attack or defence. To remove this uncertainty, a reconnoissance becomes necessary, which must be pushed on past the screen of post and sentinels, to a point where the officer in charge may see plainly the hostile army, appreciate the advantages or inconveniences of its position, count its battalions, judge of its means of resistance, whether it is intrenched and is strong in artillery, whether the ground is favorable for the action of cavalry, &c. These different objects should be seized with the promptness of a practised eye, for such an operation will usually call to arms the whole of the enemy's force, which may be very superior in numbers, and it may be necessary to leave the ground speedily But the object has been attained, since the enemy has displayed his forces.

Such reconnoissances usually precede battles. By means of them the general ascertains the true condition of affairs, before giving his final orders; he sees whether the corps of the enemy have the positions they have been reported to occupy; are in supporting distance of each other; are too much spread out; have their wings supported, &c. On the march, the advanced guard reconnoitres the enemy. Sometimes the reconnoissance is made in order to discover if a single point is strongly occupied; if a bridge has been broken; if a defile is fortified; if the enemy has cannon in a particular place; if he holds a certain village; if he is moving along such and such a road after a defeat, &c. In some cases the reconnoitring party is less numerous, and composed of cavalry alone, in order that its march may be more rapid. But when the party is expected to have an engagement, it should be composed of all arms and commanded by an officer of experience.

The commander usually receives written instructions. These he should understand perfectly before setting out. He should have a map, a glass, and all necessary materials for writing and sketching. He should procure two or three of the residents of the country to act as guides, and answer his questions relative to the names and sizes of villages, the character of the roads, the extent of woods, the features of streams and the country generally, &c. He should take with him an officer who speaks the language of the country;

it is better still if he knows the language himself. Finally, he should carefully inspect every man, animal and vehicle of the command, to be certain that all are in good condition and provided with provisions and munitions.

A detachment, intrusted with such a duty, should take all the precautions of advanced guard, flankers, &c.; scouts should examine hollow roads, woods, villages, every place where the enemy might be concealed. Every one who is met should be questioned as to the road, &c. When a detachment passes through a village, the commander should halt some minutes to obtain news of the position of the enemy. If he suspects deceit, he should take hostages, and release them only when he is certain that correct information has been given him. His inquiries should not be too prolonged, lest the enemy might learn his whereabouts, and take measures to thwart him in his object.

The commander should note particularly the features of the country he passes through, in order to determine in advance points upon which he might fall back and make a stand, if necessary. At cross-roads he should question the guides, to ascertain their direction and termination. He will test and correct his map. To get an accurate notion of the country, he should turn back frequently and examine it in its various aspects from different points of view.

The detachment remains united as well as possible

during the march, in order to be always in condition to sustain an attack or drive in an advanced post. The main body should, therefore, not lose sight of the advanced guard, nor that of its scouts. At night the scouts may be called in, as useless, or, at least, they may be drawn in within hearing distance. In such cases, frequent halts must be made in order to listen. Villages should not be entered if any thing suspicious is discovered, and until the return of two or three men sent out to examine them.

When the length of the march requires a halt, it should be made behind a wood, or some feature of the ground which will conceal the troops. Sentinels, placed in the woods or upon high points, so as to see the country without being seen, will give notice of every thing that goes on. Cavalry sentinels, in pairs, will be posted at a still greater distance on the roads towards the enemy, in order that one may leave their post and bring any information the commander should have, while the other remains on the lookout. If it is necessary to halt near a village to take food, the troops should pass through and take position on the side next the enemy, in order that he may not seize it. The commander will oblige the inhabitants to supply provisions, which he will distribute, neither officers nor men being permitted to occupy houses. A guard will be stationed in the village to enforce this regulation, and to see that the requisition for provisions is complied with. It is the better plan to pay for these, if possible.

So long as the object of the reconnoissance is not attained, the detachment should press forward, without fear of being compromised, as it is here supposed to be strong enough to overthrow easily any body of troops it may probably encounter. There are other reconnoissances, made by small detachments, where stratagem is employed, rather than force, to discover what is desired, and in such cases, of course, an engagement is to be avoided. In a strong reconnoissance, on the contrary, where the object is to penetrate as far as the positions of the enemy, the detachment should not permit itself to be checked by any body of troops it may chance to meet, but rather regard it as a piece of good fortune, as prisoners may be captured who can give valuable information, and a running fight may be kept up to the outposts of the enemy, where the flying party will give the alarm. The line is soon pierced, and a good opportunity is given of seeing the troops of the enemy as they deploy to repel the attack. They are thus forced to show their strength. The commander should, at this moment, look for a high point whence he may examine the ground and get an idea of the position itself, and the manner in which it is occupied, as well as the force and composition of the troops. He makes, or causes to be made, a rapid sketch of the ground and the position of the enemy.

He should not, in the heat of a first success, permit an inconsiderate pursuit of the fugitives, and, losing sight of the object of his expedition, become engaged so deeply as to be unable to withdraw. On the contrary, he must know how to stop and order a retreat, even in the midst of a successful engagement, when he has seen what he wishes. Prudence must here control courage, and coolness regulate the whole operation.

The commander of a reconnoitring party should avoid an engagement that would draw him off from his object, for his mission is not to damage the enemy, but to discover his plans and acquire an accurate idea of his position. When it comes in his way, he may attack, defeat detachments of the enemy, make prisoners, capture or spike cannon; but he would be reprehensible for turning out of his route and losing time to attempt to surprise a park, or carry off a convoy, or disperse a careless body of troops, even if successful. Such operations are not allowable, unless it would be too dangerous to advance without driving off troops who might be at the sides of his route.

Secret reconnoissances are conducted upon principles entirely different. Few men are taken, in order to elude better the observation of the enemy or his pursuit. The effort is made to approach by night the point that is to be reached. The detachment moves stealthily through ravines and hollow roads, making long detours to avoid meeting the patrols of the enemy, and returning by some other way, so as to escape any trap he may have laid.

A party of this sort is usually composed of but one

kind of troops, cavalry in a flat open country, and infantry in a rugged or obstructed country. It may be commanded by an officer of low rank, if he is intelligent and brave. From such services young officers acquire reputation and receive advancement. The detachment moves cautiously; it is not of sufficient numbers to have an advanced guard, but is preceded by a group of scouts, and one of these is in front of the others. Two or three flankers are necessary on each side. The commander has with him a guide or two, a special necessity for him, as he must often leave the main roads. If the scouts report the presence of a body of the enemy, he turns to one side and tries to avoid it, by riding behind a clump of trees, a rise in the ground, or in any other way. If he cannot escape observation, he will fight or retreat, according to the numbers of the enemy. If it is a strong column of the enemy, and he has been able to remain concealed, he scans it closely and endeavors to form an accurate idea of its strength, before giving notice of its presence. He then sends one of the fleetest soldiers by a détour, with a brief note to the general. In the mean time, if the column is about to reach the advanced posts, he should attack, in order to stop it and give the alarm, but he should be careful not to engage the main body; indeed, to avoid every thing but the leading detachment of the advanced guard, if the enemy is in force. His object is thus accomplished. The enemy, being uncertain what force is before him, is compelled to halt and prepare for defence. Time is thus gained. The commander of the reconnoitring party will take care to send several orderlies to the commander of the advanced posts; but, as they might be captured, and the firing might not be heard, several bundles of straw should be set on fire.

When the party has reached its destination, the commander conceals it behind a screen of some kind, as a clump of trees, a hill, an old wall, in a ravine or hollow, and, taking a few men whom he places along at intervals, he ascends to some place where he can see the enemy, being accompanied to that point only by the guide and two or three men. He makes notes of what he sees, along with the explanations given him by the guide. If the positions of the enemy are in plain sight, he makes a sketch, which it is well to do, even if the drawing is very rough. He must not permit himself to be satisfied with a mere glance, but examine every thing with coolness, and endeavor to gain exact information, even at considerable personal risk; false reports are worse than total ignorance in such matters. It requires experience to perform a duty of this kind satisfactorily. Many examples might be cited where incorrect information, made so either by inexperience or fear, has led to grave mistakes or serious disasters. If the officer is discovered while making his observations, and sees a body of troops approaching that he cannot resist with the few men near him, he hastens to rejoin the detachment at the foot of the hill, in order to fall back upon the reserve, which will advance to his support, as concealment can no longer be practised. When his whole force is united, he may attempt to repulse the enemy, if his task is not completed; but if it is, he should retire, even if sure of success. When the object of his mission is once attained, it is to his credit to bring back his whole command rather than be seeking trophies from the enemy. In this connection may be mentioned a remarkable instance of presence of mind. At the siege of Luxembourg, Vauban, wishing to ascertain, by personal observation, the real condition of affairs, as was his custom, advanced under the escort of a few grenadiers, who were left in rear lying upon the ground. He was crossing the glacis alone, under cover of the twilight darkness, when he was discovered. He beckoned with his hand to the sentinels not to fire, and continued to advance instead of retiring. The enemy took him for one of themselves, and Vauban, having seen what he wished, retired slowly and was saved by his admirable coolness.

We will indicate some of the precautions that should be taken to assure the success of a secret reconnoissance. In the first place, the detachment should consist of trusty men, and of non-commissioned officers who, in case of need, might take the place of officers. If they know the language of the enemy, it will help them greatly in attempting those ruses, by which a quick wit extricates its possessor from critical situations. In every case it is necessary for some of the party to speak the language. Two guides should be obtained, who are mounted, if the detachment is of cavalry, and always watched by men detailed for that purpose. They should not be permitted to communicate with each other, as they might concert some plan for the destruction or injury of the party. Silence should be observed, but especially at night, when every man should be all ears. When there is danger of discovery, main roads should be avoided and cross-roads used, efforts being made by the party to find concealment behind woods, hedges, rising ground; with this object, fields may be crossed; it is not safe to pass by a hill without ascertaining whether any one is behind it, a duty performed by the scouts, who make signs if all is safe. The same precaution is necessary in passing near woods, ravines, farm-houses, &c., where the enemy might be concealed. Villages are avoided as much as possible, in order not to give the alarm, especially in going out upon the expedition; if it is necessary to pass through a village, it should be previously examined with care; some of the principal persons should be questioned as to the position and plans of the enemy; at the same time, incorrect information will be given as to the destination of the detachment. Food and other necessaries will be procured for the troops, if they were not brought along. The best plan is to carry every thing that is necessary, but sometimes it is impossible to do this, for want of time; the first opportunity should then be taken to procure what is wanting.

At night the men are not permitted to smoke. When the party arrives, at daybreak, near the advanced posts of the enemy, it should be carefully put under cover, because it is the time when patrols are moving about; sentinels are placed upon all the avenues, so that notice may be given of the approach of patrols, or measures may be taken to capture them. If prisoners are taken, useful information is obtained. This is also an excellent time for ascertaining the strength of the enemy, for the troops are usually under arms until the return of the patrols. The commander will look for a high point, from which he may, by using his glass and without being seen, discover what is passing at the outposts. The fires, smoking in the distance, indicate, in a general manner, the force and location of the corps covered by the advanced posts. If it is necessary to penetrate so far, it can only be done by a night march, and by making a wide détour, in order to reach, by the rear, the villages occupied by the enemy; for there would be no probability of success in attempting to pass through the line of posts covering the front. At this time the men who speak the language are useful; they approach the villages carefully, replying to the challenges of the sentinels, and gaining access to some farm-house in order to question the inhabitants. The sentinels in the rear are generally neither numerous nor vigilant, and it is possible to capture them and learn every thing that may be desired. To accomplish such an object, the French company, which gained such renown at the siege of Dantzic, under the orders of Chambure, made use of the following stratagem: several soldiers obtained bells and, mingling with the cattle, crept up to the sentinels and killed them; they then attacked the post by assault, and the remainder of the company, which had been concealed, ran up, and the village was soon in their possession. In order to avoid recognition by their accent, Chambure's men replied in Russian to the Prussians, when they crossed the posts, and in German to the Russians. There will be many opportunities for imitating this ruse, when the hostile army is composed of troops of two or more nations.

When the reconnoissance is finished, the officer who has had charge of it gives the general a written report, when a verbal report would be insufficient. This should be clear, simple, and as brief as possible. The object of the report is the important thing, and not the more or less elegant manner in which it is written. The officer should mention only those things of which he is certain; his conjectures should be presented with caution; he should carefully avoid drawing upon his imagination for facts. Finally, he should not speak much of himself; for if there is ground for pride on account of the manner in which the duty has been performed, the troops should receive all the praise, from him at least.

The small reconnoissances every morning at the advanced posts, to ascertain that the enemy has not come nearer during the night, are made by patrols, who advance a short distance beyond the line of posts. This is a special duty for which all the officers should be detailed in turn. They seldom remain out more than an hour, and in the mean time the grand guards and pickets are kept under arms. All the precautions prescribed for secret reconnoissances are applicable to these. The commander should move with much circumspection, sending out scouts, advancing in silence, and under such covers as the country may afford, stopping often to listen and examine, giving the alarm if the enemy is met, and avoiding engagements with him. If the reconnoissance has a special object, at some distance, and it is necessary to know the result speedily, mounted men should be distributed along the route passed over, to form a continuous connection between the detachment and the outpost.

Among the means of obtaining news of the enemy, spies should be mentioned. It is, unfortunately, too true that men are found everywhere who are ready to sell their honor and to betray their country for a greater or less sum, according to the position they occupy in society. This means, costly as it is, should not be neglected, for news opportunely received often decides the result of an enterprise. Large sums may therefore be paid for such services, however contemptible the agents are. They are often in the pay of both parties, and spying for both. A treacherous spy should receive no pity if his guilt is clearly proved.

Art. II.—Topographical Reconnoissances.

Such reconnoissances are no less important than those treated above, as a general cannot arrange a plan of attack, or make the least movement of his forces, unless he has an accurate knowledge of the ground. He should know the distances between places and the obstacles to be encountered, in order to arrange a combined march of several columns. This information can only be obtained by special reconnoissances, for the most finished maps are incomplete for his uses; they never show the nature of the soil, the quality of the roads, the state of rivers and bridges, the thickness of woods, nor the exact slopes of mountains and hills, which are all things that must be known before any plan can be carried out.

It is difficult to give a perfect representation of ground upon paper; at any rate, it can only be done by those who have bestowed upon the subject much study and practice. Moreover, to make an accurate map requires time. I propose, therefore, several simplifications. Slopes may be indicated by two lines, one at the top and the other at the bottom, marking simply the contour. These are not level lines, but are readily seized by the eye, and indeed present themselves when it is desired to give but an outline of a plateau, a hill, a ridge, &c. The space between these lines gives room enough to write a few descriptive words. It will be noted whether the slope is gentle or steep, accessible or not for cavalry, what is its approximate height. In order that the *circumscribing* lines of heights be not confounded with those which give other indications, they are made broken, as shown in Figure 27. At the top of the sketch the two circumscribing lines point out plainly a plateau which is connected with the plain by a gentle slope. Lower, and near the river, there are two other lines representing an elongated hill, and what is written between the curves gives an idea of the nature of the slopes, or, at least, all that is necessary in a military point of view. On the right of the sketch is a spur shown by the circumscribing lines, terminated at the river by steep rocky slopes; a small hill, rising at its extremity, is also indicated by its circumscribing lines. The numbers in brackets give the heights of the points of the upper curve above those directly below in the lower curve, it being understood that these heights are only estimated. We see, therefore, that the plateau is on the right, 100 feet above the plain, in the centre 90, and on the left 75; the isolated hill is 10 feet high at one end, and 12 feet at the other, &c. It is not easy to judge of heights by the eye, and they are rarely placed upon a mere sketch. I have only desired to show how it is possible to put such informa-

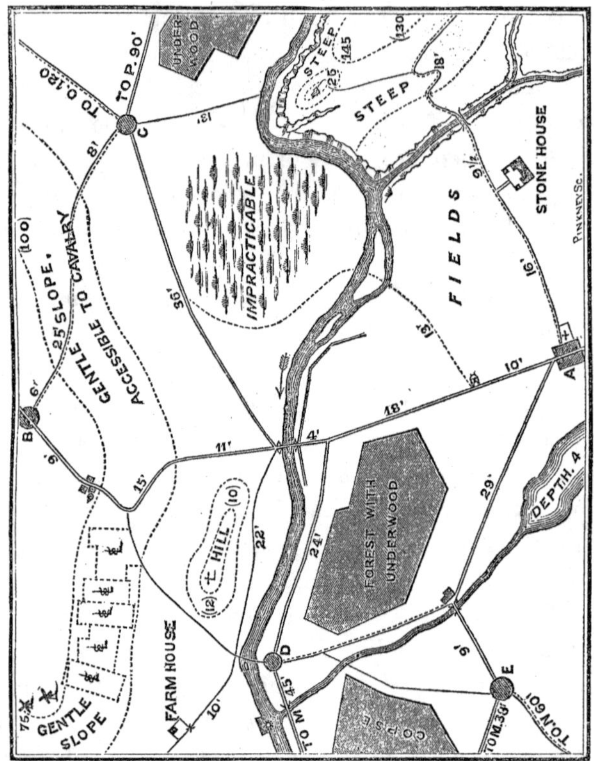

Fig. 27.

tion on a sketch without causing confusion. Every officer should be able, by using such simple means as those indicated, to represent, with considerable accuracy, the features of a piece of ground he has passed over attentively.

Having now shown how a military sketch may be

made in a simple manner, it remains to settle upon some conventional signs to represent the different objects found in a country. These are streams, ponds, marshes, roads, rocks, woods, vineyards, towns, villages, large isolated houses, which may play an important part in a combat, dikes, boats, fords, bridges of stone or wood. All of these are shown in Figure 27. We will examine them separately.

Water-courses.—These are represented either by two lines, one larger than the other, or by a succession of parallel lines between the first two. A blue tint might well replace the interior lines. The arrow indicates the direction of the current. On the lower part of the stream is a water-mill. The affluent, on the right is a brook whose banks are steep, which is indicated by two irregular lines traced along the banks.

Means of Passing.—On the left is a ferry-boat, in the centre a stone bridge, which is distinguished from the wooden bridge constructed over the affluent by the narrowness of the latter and its having no abutments on the banks. Higher up is a ford, indicated by a finely dotted line across the stream.

Stagnant Water.—Ponds and lakes are represented in the same way as rivers, the only difference being in the shape. Marshes are shown by an outline and several lines drawn parallel to the bottom of the paper with a few tufts of grass scattered. It may be noted whether the marsh is impracticable or not.

Woods and Vineyards.—It takes much time to draw these in the usual way. We have simplified it by tracing the outline, and making on the inside straight parallel marks for woods, and a small imitation of a vine for vineyards. A tint of Indian-ink or yellowish green may replace the marks for woods, if colors are used, as well as a purple tint for vineyards. Besides the marks, the character of the woods should be written.

Rocks.—The most difficult thing is the representation of rocks, on account of their great variety of form. A single sign is not sufficient to distinguish all the cases. An attempt must be made to imitate them to a certain degree. When they have long walls, it will be sufficient, as shown in the sketch, to draw the top and bottom line with a few crossings to give an idea of the object. This method is shown on the right of the drawing.

Habitations.—If it were necessary to draw all the houses in towns and villages, as in a regular topographical map, it would require considerable time, and discourage most officers. If a simple sign is used to mark the place, the thing becomes very easy. A village will, therefore, be represented by a circle crossed closely, and the town by changing the circle into a square. A is a town on our sketch; B, C, D, and E are villages. If it is important to have the form of one of those villages, a special sheet will be necessary on a large scale. A red tint may replace the cross lines.

Isolated houses, such as farm-houses, large enclosures, post-houses, inns, &c., are shown in their real form, and without regard to the scale. A small bugle or horn may indicate a post-house; a cup, an inn. On the left of the plateau is the usual sign for windmills.

Communications.—What has been said of isolated buildings applies also to roads, that is to say, it is necessary to exaggerate their width, in order to make them more visible. A main road, such as that from A to B, is represented by two parallel lines; in the same way the wagon road from C to D, with the single difference that the lines are made somewhat nearer together. Roads practicable only for small vehicles, such as that from C to B, are indicated by a single full line and a dotted line; paths for pack animals, by a single line; three of these meet near the ferry. Distances being very important in a drawing, such as we are considering, they should be written along the roads between striking objects. The numbers indicate the times required by a foot soldier to pass over the corresponding spaces, at the rate of three miles to the hour. If these distances were taken from the walk of a horse, a note should be appended, telling the rate of movement of the horse. When roads lie partly on the sheet, there should be written on each the name of the place towards which it leads, with an indication of the distance, if it is known.

Levees and Dikes.—It is important to represent

these objects, because troops may be sheltered by them. If the parallel lines were used, they might be taken to mean a road; to avoid this uncertainty, the space between the lines is filled by cross marks. This is shown upon the left bank of the stream near the bridge.

These are the different conventional signs, by means of which an officer may promptly and easily trace upon paper the result of his observations in a reconnoissance. He will not produce a fine map, but he will have a *military sketch*, which may be very useful, if the forms of objects and relative distances are only approximately obtained. To complete the sketch, a meridian line should be added, and a scale of yards, to determine distances not written. This scale is deduced from the space a man on foot can pass over in an hour, about three miles. Also, as distances along the roads are expressed in the time required to pass over them, it will be well to add a second scale agreeing with the first, giving distances passed in a minute. Officers will find it convenient to have these scales marked upon a little rule, in order to save them the trouble of re-marking them continually.

An officer who expects to make a reconnoissance, should prepare his paper, if he has time, by attaching it to pasteboard, to keep the wind from blowing it about, and by marking upon it the scales and, approximately, the positions of known objects within the limits of his work, but so lightly that they can be effaced and corrected, if necessary.

We will suppose that the patrols of the enemy have shown themselves in the country, and that the officer has been put on his guard against them. He has arrived the evening before in the town A, with his detachment; he has slept there, and gathered all the information he can with respect to the surrounding country. He knows already the population of B, C, D, and E, their distances apart, and also from the places to which roads lead from them; he learns that, in addition to the bridge on the main road, he can pass the river in the ferry-boat at D, and that there should be, higher up, a ford practicable for cavalry. He has embodied this information in his note-book, and has secured a good guide before seeking any rest. It is scarcely necessary to say that the usual military precautions for safety should be taken, as he is informed of the presence of the enemy upon the heights across the river.

The next day he should be on his way before the rising of the sun, his detachment having been assembled at dawn. A small party is sent out in charge of an officer, to examine the villages D and E, and ascertain whether any one is there; another party is sent out to the right to visit the country house, and to scout the ravine and the banks of the stream. These parties are ordered to rejoin him at the bridge. He proceeds along the main road with the principal body of the troops, but stops a half hour at the inn, to give time for the two detachments to make their circuits, and collect precise information about the ford which has been spoken of to him, some one being sent to sound it. He then moves on, marking distances by his watch, and commencing to trace the direction upon his paper. Having reached the bridge, he waits until his detachments rejoin him. A third of the troops are then left to guard the bridge, and he continues with the remainder to move on towards B, preceded by scouts and accompanied by flankers, who never should be out of sight; he sketches in the road as he advances, laying down distances by the times required to pass over them. He marks the branching of roads and paths, the feet and tops of slopes, post-houses, &c. On the subject of slopes, it should be observed that the lengths given by the scale of minutes should be diminished in proportion to the steepness of the slope, because the horizontal distances, which are alone put upon the sketch, are less in slopes; and also, in such ground less space is passed over in a given time. A reduction of one-tenth for gentle slopes is sufficient, and one-fifth for steeper. The eye, when it is practised in the estimation of distances, is a great help in such cases. Every soldier should endeavor to acquire this faculty.

Arrived at the village B, the officer leaves a third of his detachment, and with the remaining portion pushes on a mile or two, to ascertain whether any one is approaching in that direction. He may then have an engagement with the enemy. If he meets a force inferior to his own, or comes on a small post, he attacks and endeavors to capture several prisoners, who may inform him of the position of the corps further on. After pushing to the front in this way, he retires rapidly to B, and then his topographical labor really begins. He is preceded by a few scouts, and is accompanied only by a good non-commissioned officer and two or three soldiers ; the main body of his detachment remains at B, where it is established in a military manner on the side towards the enemy.

The officer goes, in the first place, to the left of the plateau, towards the windmills, in order to see the slopes and country on that side. He passes around the plateau, and returns to the village by way of the post-house. Thence he goes to the village C, with half of the troops he has with him ; the other half is ordered to leave the village in an hour, and take position at the foot of the slopes. The officer, preceded by scouts, advances along the road B C, stopping a moment at the top and bottom of the slopes to mark their direction on his sketch. In the village C he learns where the two roads go from that point, and what are the distances of the nearest villages; these will be written upon the plan. He proceeds as far as the stream by the path, and crosses the wood and returns to the village. In this way the principal bend of the river is marked upon the sketch. From the village two men should attempt to cross the marsh and arrive at the ford ; it will only be after receiving their report that the officer will write the word *impracticable*, and it would be better still, if he has the time, to ascertain the fact for himself, for subordinates are apt to be deterred by small difficulties. From C, the detachment returns to the bridge by the road. In passing, the officer detaches two men to go around the marsh, cross the ford, and return to A. He then estimates the distance to the foot of the hill, and puts in the curve on his plan.

He next passes around the little hill, following out the foot of the slopes and the paths; he ascends the hill, whence he discovers plainly the form of the river; he sketches both in, and proceeds as far as the farm-house, always measuring the distances. The farm-house is marked down, and he returns a second time to the bridge by the path, after taking note of the ferry and the mill. The whole force is now collected between the bridge and the hills, where it will remain an hour, while the commander, accompanied by the guide and four or five horsemen, passes around the villages D and E; the troops will then proceed to A and wait for their commander. The latter, in passing over this part of the ground, marks upon his sketch the woods, pond, and stream, as well as information concerning the roads from D and E.

By the time the officer reaches A, the daylight may be past, or a portion of it still left. In the latter case, he will let his troops rest, and, if he is certain of not meeting the enemy, he will take two or three of the best men and go to the heights on the right, passing by the stone country-house. From the old chateau, on the sugar-loaf at the end of the spur, he can see distinctly the course of the stream, and correct his sketch, if necessary. He will endeavor to represent this broken ground as well as he can. In returning, he will go down the little brook, down the river to the bridge, measuring the length of the dike. His work is done, and he should then return to his quarters to complete his sketch and notes.

This sketch will be undoubtedly insufficient, if an accurate map is required, for it is not to be expected that a work so rapidly done will not contain considerable errors; but what is essential is shown in a clear and simple manner, and the general can, by its aid, discover what it is important for him to know. A sketch, made with the pencil, should be gone over with pen and ink.

When the duty can be divided between several officers, a reconnoissance, even of greater extent, may be made more rapidly. One officer examines the roads; a second the streams, noting their breadth, depth, velocity, height of banks, &c.; a third goes through woods, examines villages as to their defensive properties and resources, &c. Each officer has his sheet, on which he makes his sketch and adds his notes. An officer of higher rank superintends the whole. He takes precautionary measures for the safety of his parties, and studies the general outline of the country.

Provided with his glass, he seeks elevated places, whence he can better see the ground to be examined.

However rapidly this may all be done, it is sometimes necessary to reconnoitre much more so, almost at a gallop even. All that can then be done is to take a few rough notes and put down a few hasty lines; the sketch must then be made from memory.

Itineraries.—Reconnoissances are much simplified when they have no other object than to indicate the features of a road; they are then termed *itineraries.* They are made in two ways; either by means of conventional signs, or by notes written in a table, with columns prepared in advance.

No march should be made without an officer being selected to make an itinerary. He notes the special features of the road, what is remarkable on the right and left, the breadths of defiles, the declivities of slopes, the repairs needed, &c. Distances from point to point are expressed in hours of march.

Itineraries of the first kind are made upon sheets prepared as in Figure 28.* The sheets are then united to represent the whole road, and mounted upon linen for preservation. The top of each sheet must be attached to the bottom of the next. The notes are commenced at the bottom of the sheet and carried on to the top. A straight line in the middle marks the road; squares or circles indicating inhabited places. Other lines to the right and left represent side roads,

* See also Regulations for U. S. Army, page 100

near which are written the names of the nearest places. Crooked lines mark streams crossing the road, &c. Every one may adopt such signs as he prefers.

If the sheet is too small to receive all the notes that are proper, a special note-book may be kept in addition.

Fig. 28.

&c.
Woods at 300 yards from the road.
Distance from A to D. 4h. 50′. The road is good between these points, except near the river, where are deep ruts, which must be filled for the passage of artillery.
The plateau offers a good position for defence. The front is covered by the marshy stream, the flanks rest on woods, leaving an interval of 4,000 yards.
TAVERN
35′
TOWN
D.
45′
25′
HIGHEST POINT
C.
1h10′
VILLAGE
B.
35′
40′
FARMHOUSE
PLATEAU
45′
30′
CITY
A.
&c.
Many scattered houses in the neighborhood.
Surrounded by old walls; accommodates 3,000 men.
River fordable. Stone bridge; admits 10 abreast.
From the bridge to the highest point, double teams are needed.
Accommodates 1,200 men.
Marshy stream.
Road to Y. in 4h. 30′. Practicable for wagons.
Not capable of defence. Accommodates 10,000 men.
General Remarks.
CONVENTIONAL SIGNS
Particular Remarks.

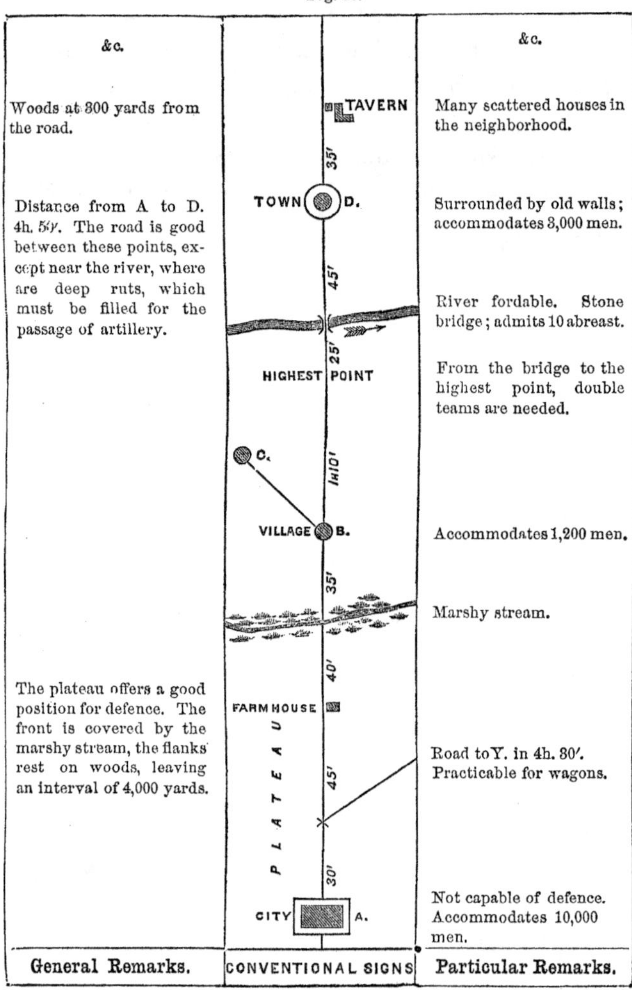

CHAPTER VIII.

SPECIAL MISSIONS, AND GIVING REST TO TROOPS.

Art. I.—Convoys.

If convoys move elsewhere than on ground controlled by the army, and, consequently, at a distance from the principal force of the enemy, they will certainly be captured, for there is no task more difficult than to defend a convoy of considerable extent against a serious attack. Frederick lost a large convoy on its way to Olmutz, because it was obliged to follow roads in the possession of the enemy. Neither the bravery of the escort nor the nearness of the army could save it from falling into the hands of the Austrians, and Frederick was obliged to raise the siege of Olmutz in consequence of its loss.

Usually these convoys are only exposed to the attacks of partisans or light troops, who have succeeded, on account of the smallness of their forces, in passing unperceived to the rear of the army. Against such attacks precautions are taken in giving escorts to convoys. These escorts are principally composed of infantry, because troops of this kind can fight on all kinds of ground, and may, in case of need, take position behind the wagons, or even in them, in order to drive off an enemy. Cavalry is also necessary, both to scout the country to a great distance from the convoy, and give prompt notice of the approach of the enemy, and also to take part in the defence against an attack of cavalry. The latter, being able to move very rapidly from one end to the other of the convoy, would soon discover some unprotected point, if the escort were composed only of foot soldiers.

It will be taken for granted, therefore, that the escort of a convoy of importance ought to be composed of infantry and cavalry, the former being the more numerous. They should march in the following order: A leading advanced guard, composed of cavalry, will precede the convoy from two to five miles, in order to examine the road in front and the ground on the flanks. But, as the enemy might escape the observation of the leading detachment, and place himself in ambush in the interval between it and the head of the convoy, a second detachment is necessary immediately in front, from which scouts will be sent out in advance and flankers on the sides. The longer is the string of carriages or pack-animals, the greater is the danger of surprise, and, consequently, the greater the necessity for precautions to avoid it. A convoy is exposed to attack in rear nearly as much as in front; it is therefore necessary to form a rear-guard, a part of which will be cavalry, to give speedy notice of what is passing in rear. The troops forming the mass of the escort, principally infantry, will be divided into three bodies, one of which will march at the head of the convoy, accompanied by workmen, carrying with them several wagons filled with tools of all kinds, ropes, timbers—indeed, every thing necessary for repairing bridges and roads. The second detachment will be placed in the middle of the column, and the third at the tail.

The troops should be by no means scattered along the whole length of the convoy, because, in case of surprise, every part would be equally weak to resist attack. There should, however, be a few men detailed from the detachments of the main body to march along the sides of the road, and compel the drivers to keep their places and distances. Order on the march is one of the most important things. The drivers are generally ready to cut loose their horses and take to flight at the first appearance of danger, and at all times to be very negligent. They should, therefore, be kept in a rigid state of discipline, and carefully watched.

The head of the convoy should move slowly and regularly, in order to avoid breaks in the column. The drivers should be forbidden to water their horses except by permission. If there is powder in the convoy, no smoking should be allowed. If a wagon breaks down, its load should be speedily divided among the others that can receive it most conveniently. There should be a signal agreed upon for the halting of the whole column when such an accident happens. If only some slight repairs are necessary, the convoy moves on, while the disabled wagon falls to one side and takes its place in rear of the column, to return to its proper place at the first general halt, unless the commander decides otherwise. The soldiers should never be permitted to put their knapsacks, &c., in the wagons; still less should they ride themselves.

Whenever the road is wide enough, the carriages should be formed in double file. The length of the column is thus diminished one-half; and, if circumstances require, the *defensive park* may be rapidly formed, which is done by the carriages wheeling to the right and left, so as to form in two compact lines, with the horses' heads towards each other and quite near. In this way the horses are better covered, the drivers cannot get off with them, and the whole space occupied is much less than when in column. This formation evidently requires some time, and should only be adopted when necessity requires.

A halt of a few minutes should be made every hour, to give the horses time to breathe, to adjust the harness, &c. In the middle of the day a longer halt is made, to feed and water the animals, but without unhitching. The necessary food for this purpose should be carried both by men and beasts. The convoy usually halts at night near a village, as provisions are to be found there for man and beast, as well as shops for repairs of harness and wagons, and for shoeing horses. Unless it is absolutely necessary to do otherwise, the park should be formed beyond the village, because it is better to have a defile of that sort in rear than in front, if it should during the night fall into the hands of the enemy.

A good place for the park is a spot enclosed by walls or thick hedges, as it is most secure. In the interior of the park the most valuable objects are put, such as despatches or other valuable papers, money, or munitions; a guard should always be placed to prevent the drivers making a fire near the wagons containing powder. The horses are tied up inside the enclosure, each team near its own wagon. The troops, except the guard of the park, bivouac at a little distance, occupying the ground, with the usual precautions of outposts, &c. No one should be permitted to leave the park without the consent of the commander. Those who are to go for provisions and forage should be designated. Neither officer, soldiers, nor drivers should be suffered to go to the inns and shops of the village. The commander should take all necessary measures for providing food for the whole command, as well as straw to sleep on; for this purpose, as well as to select a place for the park, officers and a few men should be sent in advance.

The park is commonly a hollow square, but the locality will always determine its form, which should make an enclosure, either to contain the horses and drivers, or as a kind of defensive work in case of attack. The carriages may be in file or side by side, according to the front of the square they form ; the rule being that all the poles be turned in the same direction and towards the place of destination. It is proper to double the carriages in file, so that in case of need the intervals of one row may be rapidly covered by pushing forward the carriages of the other. When the space occupied is contracted, and the number of carriages large, they are placed in several lines, with sufficient intervals between them to receive the teams.

When the convoy sets out the next day, each carriage retakes its own place in the column, and, to enable it to do so, each has its own number, which prevents all quarrels on the subject of precedence among the drivers. The officers should also see that each enters the column in its turn and without delay. These precautions may appear very minute, but they are quite necessary. There are others that should be taken by the commander of the convoy ; for example, he should satisfy himself that all the carriages and animals are in good order before setting out. He should see whether the loads are not too heavy, and remedy the difficulty if they are, either by employing more wagons or lengthening the teams. He prepares a list of all the wagons, with their contents and the names of the drivers. He requires the officers to assist in his duties of supervision, and gives them instructions what they should do in different circumstances that he may foresee. He should cause a defensive park to be formed several times, that the drivers may know how to perform their part without confusion.

The passages of defiles are dangerous for a convoy which is surrounded by numerous parties of the enemy. A defile should therefore never be entered until the outlet from it is secured, and it should be passed with all practicable speed. In passing bridges, precautions should be taken against attacks on both banks. It is manifestly the duty of a commander of a convoy to obtain all the information he can of the route he is to follow, and even to make a personal examination of it in advance, if he has time enough.

When the enemy is reported in front by the leading detachment, which falls back rapidly upon the escort, the carriages close up as much as possible and halt, or, better still, form in double file, if the road is wide enough. The advanced guard and the head detachment of the escort take a position to receive the enemy; the centre detachment forms alongside of the first, or in echelon, as the ground may permit, and as one or other flank is more exposed. The third detachment, joined by the rear-guard, remains in reserve, ready to charge the enemy when he has turned the wing of the troops in front, and is attempting to fall upon the flank of the convoy, to cut it in two, or throw it into confusion. This reserve should be immediately in front of the carriages, that it may pass readily from right to left. In this position it has the advantage of replying by small detachments to the circling movements of the attacking party. All the cavalry of the escort should be in the reserve. If the main body is too near the head of the convoy, the reserve may be obliged to take position somewhere on the flank; in this case an opening should be made in the column, to allow passage through from side to side. It is very necessary for the reserve to have much mobility, because it is called upon to look out for the feints of the enemy, and repel all the lateral attacks, which the main body cannot attend to. The soldiers placed along the convoy, to watch the drivers, must now give special attention to their duties, and should at once shoot down any who attempt to cut loose their horses in order to escape.

If the attack is repulsed, the commander of the escort must be very careful how he indulges in a pursuit, for the flight may be a feint, to draw him into an ambuscade, or away from the train, while another body of the enemy attacks it. His object is to arrive safely at his destination with his charge. This, however, should not prevent him from attacking the enemy if suitable opportunities offer. If necessity requires, he should take shelter in and behind the wagons. In a desperate case like this, a protracted resistance may be useless, on account of the great superiority of the enemy; the commander may then endeavor to save a portion of the train by abandoning the remainder, or else to destroy the whole by killing or maiming the horses, breaking the wheels, overturning the wagons, and even setting on fire those easily burned.

The defence against an attack from the rear is conducted on the same principles: — the centre detachment is united to the rear detachment and the rear-guard, to form a main body to resist the enemy; the head detachment falls back to form a reserve. The convoy should continue its march while the troops fight. The latter will retire gradually, keeping at no great distance from the tail of the column, but presenting a bold front in defiles, and wherever the ground permits. In the mean time, the advanced guard receives information of the attack, and joins the reserve.

The most dangerous attack is that in flank, because the convoy is more exposed in that direction. In this case, the three detachments unite on the side attacked, and move out so far that the enemy will be obliged to make a wide détour in attacking and expose his own flank. The best arrangement is to throw the centre detachment well out, and form in echelon the other two, strengthened respectively by the advanced and rear guard. The convoy doubles its files and moves on, regulating its movements by those of the troops. If they halt, it must halt, unless the enemy has only infantry, in which case the convoy will take the trot and escape. But the attacking party usually contains cavalry, and the convoy must regulate its movements by those of the escort. However, it may sometimes gain ground in a defile, or in some place where cavalry is unable to act. At any rate, the director of the train must obey exactly the orders of the commander of the escort. If the latter thinks it necessary to form a defensive park, the former executes the movement at the first suitable place.

It is sometimes impossible to adopt the above dispositions. There may not be men enough to form three detachments, and in this case two are used, one at the head of the convoy, and the other at the tail; perhaps, even, after furnishing the advanced and rear guards, the main body may be too small for subdivision, and then it should all march together on the most dangerous side. The rule is never to scatter the escort along the convoy, but to assemble it in one or more groups capable of effectual resistance. A convoy of importance should never be risked on the road without a previous sweeping of the country by movable columns. If the convoy is put in motion immediately after the return of the columns, there are many chances in its favor, and the escort may be considerably diminished.

It is evident, from what precedes, that the attack of a convoy is a very safe operation, even for a body of troops inferior to the escort, for, if the enemy is taken unawares, the convoy may be destroyed, or a part of it captured; if the attack fails, the party may retire safely. It should be partly infantry and partly cavalry. It is evident that if, by hiding behind a wood, a hill, a field of grain, &c., the head or tail of the convoy can be surprised and enveloped before assistance can arrive, the success is complete. This plan should be attempted before resorting to an open attack. But it is not right to expect such negligence upon the part of the commander of the escort; but, on the contrary, that the escort will be prepared for the attack. Hence, his attention must be divided by sending against him several small columns and many skirmishers, who seek to get near the wagons, to shoot the horses and obstruct the road. The cavalry, circling around, moves rapidly upon exposed points. If several carriages are reached, it is sufficient to frighten off the drivers and cut the traces, because in this way the rear part of the convoy is brought to a halt.

If the time and place of the attack can be chosen, it evidently should be made when the convoy is partly in a defile, and the head or the tail may be enveloped. Success is then certain, the inevitable crowding and confusion in the defile preventing one part of the troops from coming to the assistance of the other. But such a piece of good fortune is rare, and there are chances enough of success to justify attacking a convoy wherever it is found.

When the whole or part of a convoy has been seized, the prize should be rapidly carried off to a secure place, before the enemy can come up in superior force and recapture it. Rather than permit this, the carriages should be destroyed, and only the most valuable part of their contents preserved and carried off on the backs of the horses. Fighting is to be avoided; that is not the object of the expedition.

Art. II.—Ambuscades.

As war is now waged, and with the great number of detachments with which armies are surrounded in the present day, ambuscades are scarcely possible, except for small bodies of troops. Very broken country is particularly suited to them, from the ease with which troops may be concealed in it. This method of attacking by surprise can only succeed when the enemy is very negligent in his marches, taking none of the usual precautions; for, as soon as notice is given of an ambuscade, it has failed. In an ambuscade, the effort should be made not only to surprise the enemy, but to envelop him and cut off his retreat; for this purpose, the troops in ambush should be divided into several bodies, in order to attack on all sides at once.

When moving to a position of ambush, the body of troops should be preceded by a small advanced guard and by scouts, in the usual manner, to avoid falling into a similar trap laid by the enemy, and to seize all persons who might carry him word of the operation. This is a rule, in fact, which should never be disregarded.

Woods, hills, rocks, tall hedges, &c., are, it is true,

the most advantageous places for ambuscades, but there are others by no means useless for such purposes, such as dikes, fields of grain, or meadows of tall grass, or plains crossed by gentle undulations, particularly as the enemy is lulled into security by not suspecting any probability of an attempt at surprise on such ground. An inventive genius can make the most of circumstances of this kind, when he sees his adversary inclined to negligence and presumption. In this way Hannibal deceived Minutius. Polybius says, that between the two camps was a hill whence either could greatly annoy the other. Hannibal determined to seize it first, but suspecting that Minutius, proud of a former success, would not fail to present himself, he had recourse to a stratagem. Although the plain commanded by the hill was generally level and open, Hannibal had observed several undulations and hollows, where some men might be hid. He accordingly distributed among them, in small bodies, 500 horse and 5,000 foot. Minutius marched up to defend the hill, without perceiving the troops in ambush, who took him in flank and rear; he was completely defeated.

In preparing an ambuscade, those places are also to be sought where the enemy cannot deploy easily; where he is obliged to move in a long, narrow column; where, on account of local difficulties, disorder may be expected to occur; where the troops are separated by obstacles, &c.

At what distance from the road, followed by the enemy, should an ambŭscade be established? This question can only be solved by considering the locality, and the kind of troops to be attacked. In general, it may be said, that if the place of concealment is too near, the flankers of the enemy will discover the trap; and if it is too far off, the enemy will have time to escape. Cavalry should station itself at a greater distance than infantry, because it can get over ground more rapidly, and the noise of the horses cannot be repressed. For this reason, and because cavalry cannot act in all kinds of ground, infantry is better suited for ambuscades. However, cavalry may be used in small bodies.

The place of ambush should be entered from the sides and rear, in order to leave no tracks on the road by which the enemy might be led to suspect something. Night is the best time for moving, in order to arrive before day at the place. It is a good plan to take, at first, another road, in order to mislead the inhabitants of the country, who may be in the interest of the enemy.

A body of troops in ambush should light no fires; each soldier should remain in the place assigned to him, whether standing, sitting, or lying; he should not move, nor hold his arms in such a position as to reflect the rays of the sun. In the daytime, a portion of the troops may sleep, if they have long to wait, because the enemy is visible at a distance, and there is ample time for preparation; but at night, every man must be on the alert, to seize the favorable moment for rushing upon the enemy, when the preconcerted signal is given.

The troops being divided, as has been stated, into several bodies, with special duties to perform, each should be informed precisely what it has to do, in order to avoid confusion and to have concert in the attack. The infantry, placed as near the road as possible, fire a single volley and then rush upon the enemy, uttering loud cries at the same time. The cavalry, posted at a greater distance, make a circuit, in order to close the way both in front and rear. The party in ambush should only rise up at a given signal; it should not stir on account of a few shots, that may be accidental. The commander is the sole judge of the moment for action, and it is for him to give the signal. Too much impatience may cause the operation to miscarry.

After an engagement, the enemy may often be drawn into an ambuscade by a feigned retreat. This ruse is well known, but still it succeeds, because an enemy who believes himself victorious, and wishes to profit by his first success, does not always take all the precautions usual in an ordinary march, and, moreover, people are made presumptuous by good fortune. In 1622, Tilly was pressing Heidelberg closely. The King of Bohemia and Mansfeld passed the Rhine to succor that city. At the news of the march of the King of Bohemia, Tilly encamped near Wisloch, in a very advantageous position. To draw him from it, Mansfeld attacked, and during the engagement caused his troops to fall back, as if worsted. Tilly pursued warmly, and advanced as far as Mingelheim, where Mansfeld had placed in ambush a part of his army, and much artillery. The Bavarians, taken unawares in this way, were at once defeated; they had 2,000 men killed, and lost their baggage, cannon, and many prisoners. Heidelberg was relieved.

When information has been received of an attempt upon the part of the enemy to prepare an ambuscade, an excellent opportunity is offered to make a counterplot, for he will be surprised at the moment when he expected to cause surprise, his troops will be demoralized, and fear will give the finishing stroke. With this object, the counter-ambuscade should be arranged as near as possible to that of the enemy. Paulin, informed by deserters that Cecinna had prepared an ambuscade for him, sent a portion of his troops to station themselves in ambush near those of the enemy, while he moved up as if he knew nothing of their design. The party of Cecinna was cut to pieces, because his men lost confidence as soon as they saw Paulin's troops show themselves.

This proves that even in arranging an ambuscade, that is to say, when a commander deems himself sufficiently in control of the country to effect a surprise of his enemy, it is necessary to be on his guard, to post sentinels, and to examine the neighborhood of the position he is going to occupy. The sentinels are not only necessary for the security of the party in ambush, but also to give notice of the arrival of the enemy who is to be surprised, and to communicate any information that may be interesting to the commander. This duty should be committed to men of intelligence and experience in war; it is well even to place an officer or non-commissioned officer on the look out, with two or three men to transmit his reports.

If the patrols or sentinels see any of the scouts of the enemy, they should not challenge, but hide, or retire without noise; the least indiscretion may cause such an enterprise to fail. If the enemy discover the ambushed party, all that remains is to rush out and seize the most exposed men.

It is well to have, on the two flanks of the ambuscade, small detachments of cavalry to ride after peasants, who might see the trap, and endeavor to give notice of its existence to the enemy. The neighing of the horses is, however, dangerous, as it may lead to discovery.

It is scarcely necessary to say that, if the troops are to remain a long time concealed, the commander, before setting out upon the expedition, should see that a supply of provisions is laid in for men and horses. Having once entered the place of concealment, none should be permitted to leave it, even in disguise, for fear of arousing the suspicions of the inhabitants, and also those of the enemy.

Art. III.—Advanced Posts.

When the circumstances of the war permit the troops to remain quietly resting, they are either in camp or cantonments. Some remarks were made in a previous chapter on the subject of cantonments. It is proposed now to enter into more detail, and to treat of castrametation, or the art of choosing and laying out camps. But first, a few words will be said as to advanced posts, applicable alike to the case of troops long in a position of repose, or simply making a temporary stay at some place.

The advanced posts of a stationary body of troops perform the same duties as the scouting parties and flankers on a march. Their office is to prevent the enemy from falling upon the main body without being perceived and the alarm given. Every body of troops which is established in camp, bivouac, or cantonments, for a long time, or for a single day, or even for a few hours, should be covered by detachments whose number and strength should be proportioned to the entire force, and the extent of ground to be occupied. These detachments form the *advanced posts*. If they are well posted, and keep a good look out, the enemy cannot present himself in any direction without being seen at a considerable distance, and time is thus given for preparations for defence. If, on the contrary, they are badly placed, are too much separated, and perform their duties negligently, one of the detachments may be captured without the knowledge of its neighbors, and the danger for the main body is so much the greater because it is reposing in a false security. Outpost service deserves, therefore, the careful attention of all officers connected with it.

The nature of the ground determines the kind of troops to be used for this purpose. In an open country cavalry will be employed, but infantry chiefly in a broken country. A few horsemen should always be with the detachments, for duty as orderlies and messengers. Mounted soldiers are also necessary as vedettes on the principal roads, to give timely notice of the approach of the enemy.

The distance at which the advanced posts should be placed cannot be fixed absolutely, because it depends upon their number, the ground, and the strength of the main body. It is plain that a large body of troops, requiring some time to assemble, should have its outposts more distant than a small body, which can get under arms in a moment. Moreover, a large force can safely extend its line of outposts, because they may be made stronger and well supported.

Advanced posts are divided into *outposts* and *grand guards*. The first, as their name implies, are outside of all, the grand guards serving as centres and rallying points for them. The grand guards furnish the outposts, each sending out as many as may be necessary for a good view of the surrounding ground. In addition to these, and intermediate between them and the main body, small reserves are placed in suitable positions; they are called supports or *pickets*.

The grand guards are placed, as much as possible, in covered positions, such as low ground, behind villages or woods, on the reverse of a hill, &c. The essential thing is to keep them out of sight of the enemy, who might attempt to attack them suddenly or capture them. The outposts are still more to the front, say about 100 paces, observing the same rule of keeping out of view of the enemy. These posts, drawn from the grand guards, furnish an extensive chain of sentinels or vedettes, who are intended to see every thing that goes on. Each picket furnishes three or four grand guards, each grand guard as many outposts, and each outpost several sentinels. The strength of the outposts cannot be always and everywhere the same, being dependent on the number of sentinels to be furnished. In each outpost there should be three times as many men as it furnishes sentinels, besides the officers and non-commissioned officers.

In a wooded country, the sentinels occupy the outer skirts, the outposts the first clear places, the grand guards those more to the rear. The pickets are placed behind the forest, and always far enough so that the enemy cannot fall upon them suddenly, under cover of the trees, even if he should have succeeded in penetrating the wood. In general, woods and other covers should be viewed with suspicion. If the forest is too extensive to be left in front, large open places should be sought out for the pickets, or else abatis made to cover them. The grand guards and outposts should take similar precautions.

Communication between the posts should be kept up by frequent patrols, either to watch the sentinels, to discover any of the enemy who may slip past in the dark, to prevent desertions, to seize spies, &c. The outposts should never be separated from the grand guards by obstacles which would prevent their prompt assembling, if the enemy should attack. The same rule should as far as possible be observed between the grand guards and pickets, but sometimes cannot, on account of the greater distance. In this latter case, the obstacle should be held by an intermediate post, to keep the enemy in check and protect the retreat of the advanced troops.

The pickets may be composed of troops of all arms, but the grand guards are either cavalry or infantry. If of infantry, there should be still a few mounted men, for purposes already mentioned. The pickets, in addition to the outposts and grand guards, should have their own small posts, or, at least, their own sentinels. Generally, too many precautions cannot be taken to guard against surprises. The officer who permits himself to be surprised is entirely inexcusable; he should always take it for granted that the posts in front of him are negligent, or may be captured by the enemy, and his arrangements should be made accordingly.

When the sentinels give notice of the approach of the enemy and see him coming, they fall back upon their respective posts, then upon the grand guards; and the latter, after ascertaining the exact state of affairs, and that a serious attack is made, retire towards the pickets, skirmishing as they go. The pickets send supports to the grand guards; receive them when they come in; fight, if it is necessary to hold their positions; or fall back slowly upon the main body.

At night, the outposts are strengthened, in order to increase the number of sentinels. If the country is wooded and favors surprises, the sentinels are drawn in nearer their outposts at nightfall, in order to avoid risk of being carried off. By taking their places at the bottom of the hills, whose tops they may have occupied during the day, they have a better chance of seeing an approaching enemy.

When the same points are occupied for several days, it is a good plan to fortify them, either by throwing up earthworks or by using abatis.

No one should leave the outposts. Houses should be occupied neither by officers nor men. The men should never take off their clothes or accoutrements. Not more than half the men should sit down or rest at the same time, the remainder being on the lookout.

The outposts receive deserters, stop travellers and inhabitants coming in, and question them. Their reports are to be received with caution, as they are often false and generally inaccurate.

Frequent patrols move about among the outposts, called *defensive patrols*, to distinguish them from those which are sent to the exterior, to greater or less distances, to make some discovery. Defensive patrols are only intended to see that the sentinels are vigilant, and that the enemy does not slip in unseen between the posts; to perform this duty, a corporal and three or four men are sufficient. When a noise is heard, a patrol is sent to discover the cause, as every little thing may be an indication of the approach of the enemy. A man is sent forward to reconnoitre, who fires his piece if he sees the enemy. The corporal causes his patrol to discharge their pieces, and they all fall back towards the nearest sentinel. If the ground is very obstructed or covered, the men of the patrol are separated, in order to examine the ground more thoroughly; but usually they keep together. The patrol should follow the direction marked out for it, be perfectly silent, march slowly, halt often and listen, avoid rattling the arms, must not smoke. If a sentinel is found neglectful of his duty, the patrol halts and a man is sent to give notice at the outpost; it will only move forward when the sentinel has been replaced.

Near dawn, patrols should be sent out more frequently, because this is the time to expect the enemy. A patrol is sometimes replaced by an intelligent soldier, who makes the rounds of the sentinels, to keep them on the alert.

Art. IV.—Laying out Camps.

In establishing a camp, either tents, barracks, or branches of trees may be employed in giving shelter, depending upon the means at hand and the duration of the encampment. A camp should always be located upon military considerations, regard being also had to the health of men and animals. It should be near wood and water, but far from swamps, which breed fatal fevers. In a military point of view, a camp should overlook the surrounding country, or at least not be overlooked, and the wings should be rested upon natural obstacles. If a river flows in front, at such a distance as to permit the army to be assembled and manœuvred, so much the better. The rear should be open, and present one or more good woods, by which a retreat may be effected if necessary.

The position of a camp seems to be therefore essentially defensive. Hence come the measures for its safety; all the avenues to the camp from the front, flanks, and rear should be held by detachments; the bridges should be particularly guarded and covered by intrenchments, fords watched and defiles occupied.

Besides these detachments, which may sometimes be at a considerable distance, the camp should be immediately surrounded by guards occupying lines parallel to the front, flanks, and rear, at a distance of one or two hundred yards. To avoid surprise, the guards may construct small defensive works, which will enable them to repel a charge of cavalry, or even to resist for some time the attack of superior forces of infantry. These works may be very rapidly constructed (See Figure 29), in which the unit of measure is the yard. If the exact form shown cannot be given, at any rate a mound of earth can be thrown up, three or four feet high.

The work may be in the form of a redan or lunette, only large enough to contain the guard when under arms. A yard along the parapet is the allowance for each man. The work is closed at the gorge by a small trench and bank of earth. Sometimes abatis may be substituted for the earthwork.

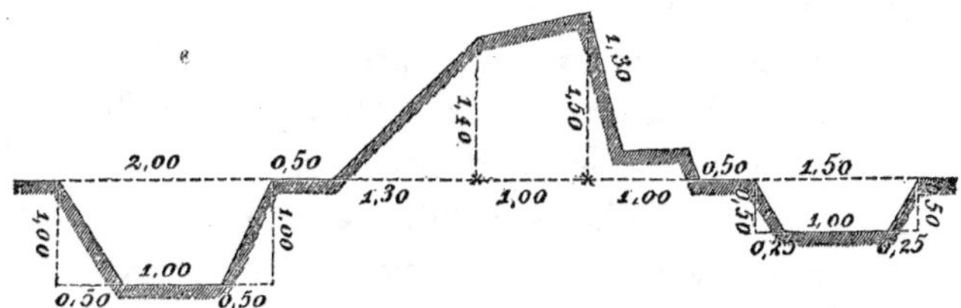

Fig. 29.

The front of the camp should be as great as that of the troops in line of battle, not only for the whole mass, but for each battalion, &c. The artillery camps are usually in rear of the infantry. As far as possible, the camp of the infantry and cavalry should form a single line, the cavalry being on the wings. The lines of the camp of an army should be regulated by the ground, in the same way as an order of battle. When the army is formed in two lines, there are two camps, one before the other. The reserve has its own. The same rules are observed in each camp, as if it were alone.

If a camp is to be occupied a long time, and may be attacked, it should be fortified. It is better to have the works that are built for its protection few and strong, than many and weak. In an open and level country, the cavalry should camp upon the wings of the infantry, because that is their place in line of battle; but in a broken country, they would form a second line behind the infantry; they should always take precautions to avoid sudden attack, as some time is required to saddle the horses, &c. The artillery park should also be in rear, as nothing is more to be feared than a surprise of a park. When there are no natural obstacles to cover the flanks of the camp, brigades or divisions may be placed in camps on the flanks, perpendicular to the general line. Sometimes, when attack may be also expected in rear, the second line may turn its color fronts to the rear, and thus a large square or rectangle be formed.

It is better, for several reasons, to put soldiers in huts than in tents: 1st, no transportation is required for them; 2d, the huts are more healthy; 3d, the construction of the huts gives occupation to the soldiers and develops their intelligence. The general arrangement of a camp of huts should be the same as for tents.

Art. V.—Bivouacs.

When the troops cannot be regularly established in camp, but still must be kept together, they are placed in bivouac. In bivouac, as in camp, each body of troops should be placed as in the order of battle.

A regiment of cavalry being in order of battle, in rear of the ground to be occupied, is broken by platoons to the right. The horses of each platoon are placed in a single row, and fastened to pickets planted firmly in the ground; near the enemy, they remain saddled all night, with girths slackened. The arms are at first stacked in rear of each row of horses; the sabres, with the bridles hung on them, are placed against the stacks. The forage is placed on the right of each row of horses. Two stable-guards for each platoon watch the horses.

A fire for each platoon is made near the color line, twenty paces to the left of the row of horses. A shelter is made for the men around the fire, if possible, and each man then places his arms and bridle against the shelter. The fires and shelter for the officers are placed in rear of the line of those for the men. The interval between the shelters should be such that the platoons can take up a line of battle freely to the front or rear.

The interval between the squadrons must be without obstruction throughout the whole depth of the bivouac. The distance from the enemy decides the manner in which the horses are fed and led to water. When it is permitted to unsaddle, the saddles are placed in the rear of the horses.

For infantry, the fires are made in rear of the color line, on the ground that would be occupied by the tents in camp. The companies are placed around them, and, if possible, construct shelters. When liable to surprise, the infantry should stand to arms at daybreak, and the cavalry mount, until the return of the reconnoitring parties. If the arms are to be taken apart to clean, it must be done by detachments, successively.

The artillery can have no fixed rule for its bivouacs, being obliged to suit itself to localities; but, in every case, the fires should not be near the park. The following is a good arrangement: 1st, a single row of fires for cannoneers and drivers; 2d, the horses in four ranks; 3d, the forage; 4th, the pieces, caissons, and wagons, parked in rear, in two ranks. The officers have a fire to the front.

Art. VI.—Cantonments.

In order to put troops in cantonments, they are distributed in the villages of the country occupied by them. When an army is moving, it must be cantoned, if at all, in the order in which it happens to be, each corps using to the best advantage the villages, hamlets, and farm-houses it occupies, all under the direction and superintendence of the officers, in order to avoid fighting among the men for the best quarters.

If the army is to remain in the same position some time, it may be more extended, in order to burden the country less and make the soldiers more comfortable. They are generally much crowded in cantoning on a march. This distribution of the troops among the villages and hamlets of a country demands much care and attention from the staff officers, who have the duty to perform, in order that the men may be securely and conveniently lodged, and that they may be uniformly distributed among the people in proportion to their ability to receive them.

Cantonments should, if possible, be established behind a river, for the sake of greater security against attack. This is particularly necessary for winter quarters, which are generally more extended. The different corps should be near enough for mutual support. A place of assembly should be designated, in view of the contingency of an attack. This point should not be too near the enemy.

The different arms of the service should be so cantoned as to afford mutual support. The artillery is placed near the roads, so as to be able to move rapidly wherever needed; it should be covered by the infantry. The cavalry may be placed wherever forage and water are most convenient, as the rapidity of its motions enables it to reach the place of assembly in time from any part of the cantonment. Care must be taken, however, not to station it where it may be easily surprised, as it is not so able to repel a sudden attack as infantry.

In each partial cantonment, there should be a place for the troops to assemble at a given signal, and all the usual precautions for safety should be taken. Each battalion should have its place of assembly in case of alarm, each brigade, each division, each army corp, and finally the whole army. This final position should be known to the commanders of divisions and even of brigades, to provide for the case of accidental separation from the divisions to which they belong.

The communication between the different places of assembly should be free; with this object, bridges should be repaired, roads improved and newly opened, if necessary. If a river passes through the cantonment, it must be bridged; and if there are deep ravines, difficult of passage, roads must be made across them for artillery. When the cantonments are not covered by an unfordable river, it is well to break up the roads by which the enemy can arrive, and to construct abatis and field-works in the most accessible places. A still better safeguard is great vigilance upon the part of the advanced posts. Both means should be used.

Generals should never leave their troops, but remain in their midst, in the most convenient place for giving their orders. In 1694, Count Tilly was captured in a house where he was lodging, because a marsh separated him from his troops. Such disgrace is reserved for officers who prefer their comfort to their safety. When troops bivouac, the generals should bivouac with them; their tents, if they have them, should be kept for maps and papers which must be used, and might be ruined by rain or dew; but they should sleep out with the men. The latter should have no ground for complaining that they are exposed to hardships and privations which are not shared by their officers. This is especially necessary with volunteers and militia.

The commanding general fixes the limits to be occupied by each corps, the commanders of corps distribute their own divisions, the generals of divisions their brigades, &c. To make these distributions, maps are required, and the more they are in detail the better. After the first assignments are made, it must be expected that many rectifications will be necessary. When they are made, tables will be prepared, showing in detail the positions of the different bodies of troops. At the head-quarters of a division, for example, there will be a table, embracing the following heads: Head-quarters of the division; Head-quarters of the brigades, battalions, isolated companies; Places of assembly of brigades; Place of assembly of the division; Remarks. So for other bodies, whether larger or smaller.

There is often no time for the preparation of such tables, where the distribution of the troops must be made very rapidly, as, for example, in cantonments on a march. In such a case, each brigade quartermaster, accompanied by the battalion quartermasters and an officer of each company, should precede the brigade several hours, to prepare lodgings, the division quartermaster having given general instructions to the brigade quartermasters. To complete the subject, there are given below some extracts from a circular addressed to commanders of divisions, in anticipation of a grand concentration of troops. It may be modified to suit varying circumstances:

1st. The companies of the same battalion must be always kept together, and, as far as practicable, the battalions and regiments of the same brigade.

2d. The soldiers of the same company must be lodged together by sections, or at least by half sections; which will not prevent putting several sections in a single house, if it affords room enough.

3d. The commander of a battalion and his staff should be placed as nearly as possible in the centre of his battalion.

4th. The company officers should be lodged in the same house with their men, or very near them.

5th. The drummers must be near head-quarters.

6th. Near the rallying place, there should always be the police-guard.

7th. Barns are to be prepared as lodgings, and the first floors of dwellings, the upper being left to the inhabitants, unless there is great scarcity of room.

8th. The company officers in advance will mark with chalk, upon the doors of barns and houses, the letters of companies, and the number of men to be received in them.

9th. When a room is six paces in width, soldiers may be made to sleep in two rows; but if the width is less, in but one row. Each man is allowed the breadth of a pace. Thus, a room twenty paces long would accommodate forty men, if six paces or more in width; but only twenty men, if less than six paces in width.

10th. The distribution of lodgings should commence on the right of the village, looking towards the enemy; that is to say, the first company should be placed on the right, and so on.

11th. The same rule should be followed if several battalions are in the same village; that is, the first battalion should be on the right.

12th. The non-commissioned officers will see that arms and accoutrements are placed in order in the several rooms, so that each man may easily lay his hands upon his own, &c., &c.

During cantonments of some duration, efforts are made to re-establish that perfect discipline and order which are often much disturbed by a long and painful campaign. During the same period, the cavalry is remounted, the ranks filled up, the troops paid, clothed, shod, and fully equipped in every way. In marches in their own country, the troops live on what they carry with them, and what may be procured for distribution to them by the proper officers. In an enemy's country, they may live on the product of requisitions made upon the inhabitants. Those requisitions should be regulated in amount by the ability of the people to meet them and by the wants of the troops, both present and prospective; for it is always well to have a supply of provisions for several days, in order that the troops may not want the necessaries of life. These requisitions are renewed as the army advances. But it is one of the most difficult things to prevent the soldiers from wasting their rations, when they have a supply issued for several days at once.

Dépôts of supplies are also formed, from which the troops may be subsisted during the operations of the campaign, if the country through which they pass cannot support them. If extended requisitions upon the country do not furnish sufficient supplies, more must be purchased elsewhere, for money must be expended to feed the men, who have, at best, to undergo many privations and sufferings.

If the war is conducted upon proper principles, the soldiers should never be permitted to seize food or clothing for themselves. Every thing should come to them through authorized officers, otherwise the war will soon degenerate into wholesale pillaging. Undoubtedly, the men must not be permitted to starve nor to suffer in an abundant country. The enemy should be always made to bear, as far as possible, the burden of the war. But, on the other hand, leaving out of consideration the calls of humanity, self-interest requires that the non-combatant inhabitants of any country be treated in person and property with as much consideration as possible. By requiring an invaded country to furnish whatever supplies it can, without causing starvation to innocent non-combatants; by exhibiting all the kindness and humanity possible, consistently with firmness and justice, provision is made for the future, and especially for the eventuality of a retrograde movement through the same country. By treating the inhabitants with justice and humanity, and giving them an equivalent for their property, many friends are gained, or, at least, their hostility diminished; they will not fly at your approach; they will give you bread for money, and give you shelter in their habitations. Act otherwise;

pillage, ravage the country; force the timid non-combatants to flee for their lives; march with the sword in one hand and the fire-brand in the other; surround yourself with ruins, make the country a desert, and in a little while you may be yourself reduced to the direst extremities. To-day you may be swimming in plenty, and to-morrow enduring the pangs of want. I repeat it; require the inhabitants of a country to supply your necessities, for this is the right of war; but use this right with wise discretion. In this light must we interpret the maxim of Cato — *War supports war.*

THE END.

TABLE OF CONTENTS.

CHAPTER I.

PRINCIPLES OF STRATEGY.

CHAPTER II.

ORGANIZATION OF ARMIES

CHAPTER III.

MARCHES AND MANŒUVRES.

CHAPTER IV.

BATTLES.

CHAPTER V.

DEFENCE OF RIVERS AND MOUNTAINS—COVERING A SIEGE.

CHAPTER VI.

COMBATS AND AFFAIRS.

CHAPTER VII.

RECONNOISSANCES.

CHAPTER VIII.

SPECIAL MISSIONS, AND GIVING REST TO TROOPS.

ARMY OFFICERS' POCKET COMPANION.

Principally designed for Staff Officers in the Field. Partly translated from the French of M. DE ROUVRE, Lieutenant-Colonel of the French Staff Corps, with Additions from standard American, French, and English Authorities. By WM. P. CRAIGHILL, First Lieutenant U. S. Corps of Engineers, Assist. Prof. of Engineering at the U. S. Military Academy, West Point. 1 vol., 18mo, full roan. $1 50.

"I have carefully examined Capt. CRAIGHILL'S Pocket Companion. I find it one of the very best works of the kind I have ever seen. Any Army or Volunteer officer who will make himself acquainted with the contents of this little book, will seldom be ignorant of his duties in camp or field."

H. W. HALLECK, Major-General U. S. A.

"I have carefully examined the 'Manual for Staff Officers in the Field.' It is a most invaluable work, admirable in arrangement, perspicuously written, abounding in most useful matters, and such a book as should be the constant pocket companion of every army officer, Regular and Volunteer."

G. W. CULLUM, Brigadier-General U. S. A.
Chief of General Halleck's Staff, Chief Engineer Department Mississippi.

MANUAL FOR ENGINEER TROOPS.

Consisting of

Part I. Ponton Drill.
II. Practical Operations of a Siege.
III. School of the Sap.
IV. Military Mining.
V. Construction of Batteries.

By Major J. C. DUANE, Corps of Engineers, U. S. Army. 1 vol., 12mo, half morocco, with plates. $2.

"I have carefully examined Capt. J. C. Duane's 'Manual for Engineer Troops,' and do not hesitate to pronounce it the very best work on the subject of which it treats."

H. W. HALLECK, Major-General, U. S. A.

"A work of this kind has been much needed in our military literature. For the army's sake, I hope the book will have a wide circulation among its officers."

G. B. MCCLELLAN, Major-General, U. S. A.

VIELE'S HAND-BOOK.

Hand-Book for Active Service, containing Practical Instructions in Campaign Duties. For the use of Volunteers. By Brig.-Gen. EGBERT L. VIELE, U. S. A. 12mo, cloth. $1.

BENÉT'S MILITARY LAW.

A Treatise on Military Law and the Practice of Courts-Martial, by Capt. S. V. BENÉT, Ordnance Department, U. S. A., late Assistant Professor of Ethics, Law, &c., Military Academy, West Point. 1 vol., 8vo, law sheep. $3.

"This book is manifestly well timed just at this particular period, and it is, without doubt, quite as happily adapted to the purpose for which it was written. It is arranged with admirable method, and written with such perspicuity and in a style so easy and graceful, as to engage the attention of every reader who may be so fortunate as to open its pages. This treatise will make a valuable addition to the library of the lawyer or the civilian; while to the military man it seems to be indispensable."—*Philadelphia Evening Journal.*

"Captain Benét presents the army with a complete compilation of the precedents and decisions of rare value which have accumulated since the creation of the office of Judge-Advocate, thoroughly digested and judiciously arranged, with an index of the most minute accuracy. Military Law and Courts-Martial are treated from the composition of the latter to the Finding and Sentence, with the Revision and Execution of the same, all set forth in a clear, exhaustive style that is a cardinal excellence in every work of legal reference. That portion of the work devoted to Evidence is especially good. In fact, the whole performance entitles the author to the thanks of the entire army, not a leading officer of which should fail to supply himself at once with so serviceable a guide to the intricacies of legal military government."—*N. Y. Times.*

"This volume has been carefully prepared by the author in order to present in a collected form, the able decisions issued from the War Department since the establishment of the office of Judge-Advocate of the Army, and the opinions given by the Attorneys-General on points requiring legal interpretation. It is full of information for military men, founded on the authority of these decisions and opinions, and is a valuable contribution to the science of military law."—*Providence Journal.*

JUDGE-ADVOCATE GENERAL'S OFFICE,
October 13, 1862.

* * * So far as I have been enabled to examine this volume, it seems to me carefully and accurately prepared, and I am satisfied that you have rendered an acceptable service to the army and the country by its publication at this moment. In consequence of the gigantic proportions so suddenly assumed by the military operations of the Government, there have been necessarily called into the field, from civil life, a vast number of officers, unacquainted from their previous studies and pursuits, both with the principles of military law and with the course of judicial proceedings under it. To all such, this treatise will prove an easily accessible storehouse of knowledge, which it is equally the duty of the soldier in command to acquire, as it is to learn to draw his sword against the common enemy. The military spirit of our people now being thoroughly aroused, added to a growing conviction that in future we may have to depend quite as much upon the bayonet as upon the ballot-box for the preservation of our institutions, cannot fail to secure to this work an extended and earnest appreciation. In bringing the results of legislation and of decisions upon the questions down to so recent a period, the author has added greatly to the interest and usefulness of the volume. Very respectfully,

Your obedient servant, J. HOLT.

ELEMENTS OF MILITARY ART AND HISTORY.

By EDWARD DE LA BARRE DUPARCQ, Chef de Bataillon of Engineers in the Army of France; and Professor of the Military Art in the Imperial School of St. Cyr. Translated by Brig-Gen. GEO. W. CULLUM, U. S. A., Chief of the Staff of Major-Gen. H. W. HALLECK, General-in-Chief U. S. Army. 1 vol., octavo, cloth. $4.

"I read the original a few years since, and considered it the very best work I had seen upon the subject. Gen. Cullum's ability and familiarity with the technical language of French military writers, are a sufficient guarantee of the correctness of his translation."
H. W. HALLECK, Major-Gen. U. S. A.

"I have read the book with great interest, and trust that it will have a large circulation. It cannot fail to do good by spreading that very knowledge, the want of which among our new, inexperienced, and untaught soldiers, has cost us so many lives, and so much toil and treasure."
M. C. MEIGS, Quartermaster-General, U. S. A.

"I have carefully read most of General Cullum's translation of M. Barre Duparcq's 'Elements of Military Art and History.' It is a plain, concise work, well suited to our service. Our volunteers should read and study it. I wish it could be widely circulated among our officers. It would give them a comprehensive knowledge of the different arms of the service, and invite further investigation into the profession of arms which they have adopted. A careful study of such works will make our officers learned and skilful, as well as wise and successful; and they have ample time, while they are campaigning, to improve themselves in this regard."
S. R. CURTIS, Major-Gen. U. S. A.

"Barre Duparcq is one of the most favorably known among recent military writers in France. If not the very best, this is certainly among the best of the numerous volumes devoted to this topic. Could this book be put into the hands and heads of our numerous intelligent, but untrained officers, it would work a transformation supremely needed. We can say, that no officer can read this work without positive advantage, and real progress as a soldier. Gen. Cullum is well known as one of the most proficient students of military science and art in our service, and is amply qualified to prepare an original text-book on this subject."—*Atlantic Monthly.*

"That Gen. Cullum, Chief of Staff to the Commanding General, should have chosen to translate and edit his work, rather than to prepare an original one himself, gives the highest professional testimony to its value. The work contains a History of the Art of War, as it has grown up from the earliest ages; describes the various formations which have from time to time been adopted; and treats in detail of the several arms of the service, and the most effective manner of employing them for offensive and defensive purposes. It is fully illustrated with diagrams, displaying to the eye the formations and evolutions which find place in ancient and modern armies. Though the book is especially designed for the instruction of officers and soldiers, the non-professional reader cannot fail to perceive the clearness of its statements and the precision of its definitions."—*Harpers' Monthly.*

RIFLES AND RIFLE PRACTICE.

An Elementary Treatise on the Theory of Rifle Firing; explaining the causes of Inaccuracy of Fire and the manner of correcting it; with descriptions of the Infantry Rifles of Europe and the United States, their Balls and Cartridges. By Capt. C. M. WILCOX, U. S. A. New edition, with engravings and cuts. Green cloth. $1.75.

"Although eminently a scientific work, special care seems to have been taken to avoid the use of technical terms, and to make the whole subject readily comprehensible to the practical enquirer. It was designed chiefly for the use of Volunteers and the Militia: but the War Department has evinced its approval of its merits by ordering from the publisher one thousand copies for the use of the United States Army."—*Louisville Journal.*

"The book will be found intensely interesting to all who are watching the changes in the art of war arising from the introduction of the new rifled arms. We recommend to our readers to buy the book."—*Military Gazette.*

"A most valuable treatise."—*New York Herald.*

"This book is quite original in its character. That character is completeness. It renders a study of most of the works on the rifle that have been published quite unnecessary, We cordially recommend the book."—*United Service Gazette, London.*

"The work being in all its parts derived from the best sources, is of the highest authority, and will be accepted as the standard on the subject of which it treats."—*New Yorker.*

NEW BAYONET EXERCISE.

A New Manual of the Bayonet, for the Army and Militia of the United States. By Colonel J. C. KELTON, U. S. A. With thirty beautifully-engraved plates. Red cloth. $1.75.

This Manual was prepared for the use of the Corps of Cadets, and has been introduced at the Military Academy with satisfactory results. It is simply the theory of the attack and defence of the sword applied to the bayonet, on the authority of men skilled in the use of arms.

The Manual contains practical lessons in Fencing, and prescribes the defence against Cavalry, and the manner of conducting a contest with a Swordsman.

"This work merits a favorable reception at the hands of all military men. It contains all the instruction necessary to enable an officer to drill his men in the use of this weapon. The introduction of the Sabre Bayonet in our army renders a knowledge of the exercise more imperative."—*New York Times.*

NEW MANUAL OF SWORD AND SABRE EXERCISE.

By Colonel J. C. KELTON, U. S. A. Thirty plates. *In press.*

SYSTEMS OF MILITARY BRIDGES,

In Use by the United States Army; those adopted by the Great European Powers; and such as are employed in British India. With Directions for the Preservation, Destruction, and Re-establishment of Bridges. By Brig.-General GEORGE W. CULLUM, Lieut.-Col. Corps of Engineers, United States Army. 1 vol., octavo. With numerous Illustrations. $3.50.

"It is a trite remark that of all the operations of war none is more difficult and hazardous than the passage of a large river in the presence of a bold and active enemy. The importance to this country of such a work as the present, when our armies have to pass so many great rivers, cannot be over-estimated. We have no man more competent to prepare such a work than Brigadier-General Cullum, who had the almost exclusive supervision, devising, building, and preparing for service of the various bridge-trains sent to our armies in Mexico during our war with that country. The treatise before us is very complete, and has evidently been prepared with scrupulous care. The descriptions of the various systems of military bridges adopted by nearly all civilized nations are very interesting even to the non-professional reader, and to those specially interested in such subjects must be very instructive, for they are evidently the work of a master of the art of military bridge-building."—*Washington Chronicle.*

MILITARY AND POLITICAL LIFE OF THE EMPEROR NAPOLEON.

By BARON JOMINI, General-in-Chief and Aide-de-camp to the Emperor of Russia. Translated from the French, with notes, by H. W. HALLECK, LL. D., Major-General U. S. Army. 4 vols., Royal octavo. Fully Illustrated by Maps and Plans. *In press*

DELAFIELD'S REPORT.

Report on the Art of War in Europe in 1854, 1855, and 1856. By Col. R. DELAFIELD, Corps of Engineers U. S. A. 1 vol. folio, cloth. With maps and views. $5.

SCOTT'S MILITARY DICTIONARY.

Comprising Technical Definitions; Information on Raising and Keeping Troops; Actual service, including makeshifts and improved *materiel*, and Law, Government, Regulation, and Administration relating to Land Forces. By Colonel H. L. SCOTT, Inspector-General U. S. A. 1 vol., large octavo, fully illustrated, half morocco, $5; Half russia, $6.50; Full morocco, $8.

"It is a complete Encyclopædia of Military Science, and fully explains every thing discovered in the art of war up to the present time."—*Philadelphia Evening Bulletin.*

"We cannot speak too much in legitimate praise of this work."—*National Intelligencer.*

"It should be made a text-book for the study of every volunteer."—*Harper's Magazine.*

"We cordially commend it to public favor."—*Washington Globe.*

"This comprehensive and skilfully prepared work supplies a want that has long been felt, and will be peculiarly valuable at this time as a book of reference."—*Boston Commercial Bulletin.*

"The Military Dictionary is splendidly got up in every way, and reflects credit on the publisher. The officers of every company in the service should possess it."—*N. Y. Tablet.*

"The work is more properly a Military Encyclopædia, and is profusely illustrated with engravings. It appears to contain every thing that can be wanted in the shape of information by officers of all grades."—*Philadelphia North American.*

"This book is really an Encyclopædia, both elementary and technical, and as such occupies a gap in military literature which has long been most inconveniently vacant. This book meets a present popular want, and will be secured not only by those embarking in the profession but by a great number of civilians, who are determined to follow the descriptions and to understand the philosophy of the various movements of the campaign. Indeed, no tolerably good library would be complete without the work."—*New York Times.*

"The work has evidently been compiled from a careful consultation of the best authorities, enriched with the results of the experience and personal knowledge of the author."—*N. Y. Daily Tribune.*

"Works like the present are invaluable. The officers of our Volunteer service would all do well to possess themselves of the volume."—*N. Y. Herald.*

"It is a book to be referred to on the spur of the moment, to be consulted at leisure, and to be read with deliberation. It reflects honor on the military service of the United States, and gives new glory to one of the noblest of the names connected with that service."—*Boston Traveller.*

THE ARTILLERIST'S MANUAL:

Compiled from various Sources, and adapted to the Service of the United States. Profusely illustrated with woodcuts and engravings on stone. Second edition, revised and corrected, with valuable additions. By Gen. JOHN GIBBON, U. S. Army. 1 vol., 8vo, half roan. $5.

This book is now considered the standard authority for that particular branch of the Service in the United States Army. The War Department, at Washington, has exhibited its thorough appreciation of the merits of this volume, the want of which has been hitherto much felt in the service, by subscribing for 700 copies.

"It is with great pleasure that we welcome the appearance of a new work on this subject, entitled 'The Artillerist's Manual,' by Capt. John Gibbon, a highly scientific and meritorious officer of artillery in our regular service. The work, an octavo volume of 500 pages, in large, clear type, appears to be well adapted to supply just what has been heretofore needed to fill the gap between the simple Manual and the more abstruse demonstrations of the science of gunnery. The whole work is profusely illustrated with woodcuts and engravings on stone, tending to give a more complete and exact idea of the various matters described in the text. The book may well be considered as a valuable and important addition to the military science of the country."—*New York Herald.*

HAND-BOOK OF ARTILLERY,

For the Service of the United States Army and Militia. New, revised, and greatly enlarged edition. By Maj. JOSEPH ROBERTS, U. S. A. 1 vol., 18mo, cloth. $1.

"A complete catechism of gun practice, covering the whole ground of this branch of military science, and adapted to militia and volunteer drill, as well as to the regular army. It has the merit of precise detail, even to the technical names of all parts of a gun, and how the smallest operations connected with its use can be best performed. It has evidently been prepared with great care, and with strict scientific accuracy. By the recommendation of a committee appointed by the commanding officer of the Artillery School at Fort Monroe, Va., it has been substituted for 'Burns' Question and Answers,' an English work which has heretofore been the text-book of instruction in this country."—*New York Century.*

The following is an extract from a Report made by the committee appointed at a meeting of the staff of the Artillery School at Fort Monroe, Va., to whom the commanding officer of the School had referred this work:

* * * "In the opinion of your Committee, the arrangement of the subjects and the selection of the several questions and answers have been judicious. The work is one which may be advantageously used for reference by the officers, and is admirably adapted to the instruction of non-commissioned officers and privates of Artillery.

"Your Committee do, therefore, recommend that it be substituted as a text-book in place of 'Burns' Questions and Answers on Artillery.'"

(Signed,) I. VOGDES, *Capt. 1st Artillery.*
(Signed,) E. O. C. ORD, *Capt. 3d Artillery.*
(Signed,) J. A. HASKIN, *Bvt. Maj. and Capt. 1st Artillery.*

CAVALRY; ITS HISTORY, MANAGEMENT, AND USES IN WAR.

By J. ROEMER, LL. D., late an Officer of Cavalry in the Service of the Netherlands. Elegantly illustrated, with one hundred and twenty-seven fine Wood Engravings. In one large octavo volume, beautifully printed on tinted paper. Price $5.00.

SUMMARY OF CONTENTS.—Cavalry in European Armies; Proportion of Cavalry to Infantry; What kind of Cavalry desirable; Cavalry Indispensable in War; Strategy and Tactics; Organization of an Army; Route Marches; Rifled Fire-Arms; The Charge; The Attack; Cavalry versus Cavalry; Cavalry versus Infantry; Cavalry versus Artillery; Field Service; Different Objects of Cavalry; Historical Sketches of Cavalry among the early Greeks, the Romans, the Middle Ages; Different kinds of Modern Cavalry; Soldiers and Officers; Various systems of Training of Cavalry Horses; Remounting; Shoeing; Veterinary Surgeons, Saddlery, etc., etc.

WHAT GENERAL M'CLELLAN SAYS OF IT.

"I am exceedingly pleased with it, and regard it as a very valuable addition to our military literature. It will certainly be regarded as a standard work, and I know of none so valuable to our cavalry officers. Its usefulness, however, is not confined to officers of cavalry alone, but it contains a great deal of general information valuable to officers of the other arms of service, especially those of the Staff.

"With the hope that it may find a large circulation in our armies, I am, very truly yours

"GEO. B. MCCLELLAN, Maj-Gen. U. S. A."

"We can truly say that, from no single volume have we obtained so much information on the art of war as from this."—*Phila. Press.*

"Reading, culture, and practical experience united, give this work a value and importance that will cause it to be regarded as a standard authority."—*Saturday Eve. Gazette.*

"By far the best treatise upon Cavalry and its uses in the field, which has yet been published in this country, for the general use of officers of all ranks, is this elaborate and interesting work. Eschewing the elementary principles and tactics of cavalry, which may be learned from any hand-book, the author treats of the uses of cavalry in the field, of strategy and tactics, and of its general discipline and management. The range of the work includes an admirable treatise upon rifled fire-arms, a historical sketch of cavalry, embodying many interesting facts, an account of the cavalry service in Europe and this country, and a treatise on horses, their equipment, management, &c. The work is copiously illustrated and elegantly printed. It is interesting not alone to military men but to the general reader, who will gain from its pages valuable historical facts and very clear ideas of some branches of the art of war, such as the employment of spies, gaining information in an enemy's country, advance movements and other strategical maneuvres.—*Boston Journal.*"

www.ingramcontent.com/pod-product-compliance
Lightning Source LLC
LaVergne TN
LVHW020106110826
845151LV00001B/77

9781425544010